From Digital Twins to Digital Selves and Beyond

Franz Barachini • Christian Stary

From Digital Twins to Digital Selves and Beyond

Engineering and Social Models for a Trans-humanist World

 Springer

Franz Barachini
Langenzersdorf, Austria

Christian Stary
JKU
Linz, Austria

ISBN 978-3-030-96411-5   ISBN 978-3-030-96412-2   (eBook)
https://doi.org/10.1007/978-3-030-96412-2

This Springer imprint is published by the registered company Springer Nature Switzerland AG
The registered company address is: Gewerbestrasse 11, 6330 Cham, Switzerland

Altering man's bodily functions to meet the requirements of extraterrestrial environments would be more logical than providing an earthly environment for him in space . . . Artifact-organism systems which would extend man's unconscious, self-regulatory controls are one possibility. (Manfred E. Clynes & Nathan S. Kline, 1960)[1]

Another possibility is to extend cyber-physical artefacts by means of cognitive computing, i.e. to design autonomous socio-emotional agents embodied in ecosystems as socially informed entities. And this is what the book is about. (The authors, 2021)

---

[1]Clynes, M. E., & Kline, N. S. (1960). Cyborgs and space. *Astronautics, 14*(9), 26–27.

To our wives, Elisabeth and Edith

Prologue

According to evolutionary theory, the first amino acids evolved from a primordial soup based on inorganic material. They formed the basis for organic life. Since then, several transformation processes of biological life were traversed. Most scientists believe that these transformation processes are not continuous but volatile. Nobody knows when exactly self-awareness evolved. Mankind evolved not steadily but gradually with sudden boosts and more steady phases in between.

Approximately 70,000 years ago, the cognitive revolution started with the development of a language. In contrast to animals, the modern homo sapiens was able to use this language as a means for storytelling. By inventing stories, myth could be created. Yuval Noah Harari (2019) coined the term virtual language for that. This virtual language allowed the creation of common imaginations. Common imaginations and myth formed the cornerstone for the management of large groups and even complete nations.

Concerning the younger history, Karl Jaspers (2010) introduced the term *axial ages* for extraordinary leaps in mental evolution. During the first axial age between 800 and 200 before Christ, the thinking frames of all higher developed societies were created within a few generations independent from each other. The first woodblock printings were also invented during that period. Later, around 1700 anno Domini, the search for a proof of God's existence yielded the foundation of natural sciences.

Since the invention of the Turing machine in the year 1936, more and more innovations and inventions were produced in shorter time. The usage of mass products follows exponential curves. This is due to the existence of modern computer systems.

In the middle of social, environmental, and technical upheavals, it is legitimate to ask in which directions humankind will evolve. Overpopulation and as a consequence environmental pollution, loss of biodiversity, and, as a consequence, pandemics and climate change threaten our existence. In the long run, global warming could make organic life on earth extinct. In order to extend human existence, mankind has some possibilities to survive for some time. We either can leave the solar system or we can try to transform ourselves back to inorganic material.

Leaving the solar system is only possible with power unit technologies which are not invented yet. We think that the transformation from inorganic material to organic life and back again is also promising. The latter step is known as *trans-humanism*. Ray Kurzweil (2005) introduced the idea of "uploading" brains on powerful computational substrates. This process and the amalgamation of artificial and real brains are known as *singularity*.

This book is a tiny contribution to the ideas of the *Gilgamesh project*. However, in contrast to the *Gilgamesh project*, which is more oriented toward the exploration of eternal biological life, we think that research initiatives redirecting attention toward capabilities of digital artifacts and their impact on artificial life help redefining common grounds and disparities. In this way, we want to lay ground for informed co-existence and co-creation opportunities, either driven by biological or digital capabilities. Consequently, the book is about deepening the understanding of the relation between Cyber-Physical Systems (CPS) as socio-technical systems and advancing their digital representations in terms of intertwining human and machine intelligence. In exploring this avenue, it is crucial to recognize that digital twins as digital selves should be able to develop emotional behavior, enriching the socio-cognitive habitat of Cyber-Physical Systems.

Why This Book?

Picture a traffic ecosystem including fully autonomous cars among other vehicles. It represents a socio-technical system as some of the cars are driven by humans and some of them are not. Fully autonomous cars do not require human drivers. These vehicles are capable of sensing their environment and are preprogrammed to operate without human intervention. As they do not require human interaction, they are technology artifacts of the socio-technical traffic ecosystem.

From a traffic system perspective, both fully autonomous cars and human-driven cars are equal components. Being quite different with respect to perception, decision-making, and operation control processes, each party has to control itself in traffic and to meet regulatory and mobility requirements. Human-driven cars, semi-automated vehicles, and fully autonomous cars can go anywhere traditionally operated cars can move and should be able to perform tasks experienced human drivers can accomplish.

Autonomous cars have some particularities human-driven ones may not have and do not rely on when it comes to socio-emotional behavior. Technologically, fully autonomous cars need to rely on sensors, actuators, complex algorithms, machine learning systems, and powerful processors to monitor behavior, identify events, plan the next intervention, and run software for behavior execution. Many human sensing capabilities are copied by autonomous cars when creating and maintaining a map of their surroundings. Radar sensors monitor the position of nearby vehicles. Video cameras detect traffic lights, read road signs, track other vehicles, and look for pedestrians. Lidar (light detection and ranging) sensors bounce pulses of light off the car's surroundings to measure distances, detect road edges, and identify lane markings. Ultrasonic sensors in the wheels detect curbs and other vehicles when parking. Overall, many detectors of environment signals collect data to be processed by technologically relevant (re-)actions.

What if digital processes do not only process sensory inputs or plot a path and send instructions to the car's actuators, which control acceleration, braking, and steering, but operate beyond these capabilities? What if a fully autonomous car does not only rely on rules avoiding obstacles, predicting driving behavior, and recognizing various objects, but rather uses algorithms for predictive modeling and situation recognition to take into account social and emotional patterns of behavior? What if cognitive computing is used to enrich cyber-physical ecosystems beyond following traffic rules and obstacle navigation and develop individual emotions?

Once socially enriched fully autonomous cars become part of the traffic system, socio-cognitive intelligence qualifies them as integrative and socially adaptive parties of this ecosystem. Similar to human-driven cars, autonomous vehicles could include behavior comparable to subtle cues and non-verbal communication. For instance, instead of making direct eye contact with other ecosystem, participants or systems expression of affordance could be encoded relating to politeness. Similar to reading facial expressions and body language of other participants, behavior judgment and activity predictions could become part of autonomous vehicle behavior.

As can be shown by developing real-life scenarios, these mechanisms target connecting to people. However, expecting only for upcoming car generations more than 300 million lines of code (cf. http://www.synopsys.com), digital modeling of autonomous systems becomes an indispensable asset in socio-technical development. Hence, we need to think about encoding socio-emotional intelligence in digital twins and establish them as digital selves. Digital selves are expected to contribute to what could termed "social compatibility" of actors in a continuously transforming digitized world. Key are transparent and executable models, structuring cognitive and social processes for humans kept in the development loop.

Overview

Since digital twins increasingly become high-fidelity digital models of ecosystems and their dynamic evolvement (cf. Durão et al., 2018) in the first part of this book, we provide information on their background and their development in the context of Cyber-Physical Systems (CPS) and their industrial application. We discuss the engineering of so-called humanoid socio-technical settings mimicking cognitive and social skills. This tendency lays ground to build up sovereignty on developing some digital identity or self, finally controlling individual development by digital means. Taking this role, digital twins are real-time models capable of controlling their adaptation and, finally, next-generation development, thus representing digital selves. They become the core of socio-trans-human developments.

In the second part of this book, we introduce relevant approaches to awareness and the emergence of consciousness in artificial agents. We start with discussing the awareness of identity in digital twins as prerequisite to social representations. We then discuss whether social behavior of digital selves can result as an emergent phenomenon. Progress in simulating emotions by using game theoretical approaches is presented.

These findings lead finally to considering the coexistence in terms of a symbiosis of digital twins and models of socio-emotional intelligence (cf. Küster et al., 2020) enabling digital selves on the background of highly connected ecosystems and increasingly digitized service operations. In order to structure the contributions from the various fields, we adopt a tri-partite framework. It has been developed to capture human-related issues, organizational elements, and technologies including their mutual relations. We adopt this well-established conceptualization for complex system analysis and development—cf. Oppl and Stary (2019) for the digitalization of work (processes) and Ackerman et al. (2018) for the socio-technical transformation of society-relevant systems, such as healthcare.

The adoption of the framework extends the human perspective to individuals in the realm of trans-human developments enabled by technology. The technology

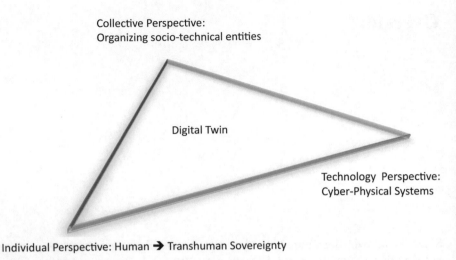

Collective Perspective:
Organizing socio-technical entities

Digital Twin

Technology Perspective:
Cyber-Physical Systems

Individual Perspective: Human ➜ Transhuman Sovereignty

Fig. 1 Digital twin development as socio-technical endeavor

perspective recognizes the cyber-physical nature of upcoming ecosystems. The collective perspective includes the socio-technical structuring of individual elements toward a systemic whole (Fig. 1).

In the following, we provide a chapter-by-chapter overview for each part of the book. The first chapter describes major milestones in computer science. These milestones form the basis for the implementation of digital twins and digital selves—the third axial age, still being in progress, started around the 1930s.

Although the idea of the digital twin stems from industrial automation, its objectives have systemic relevance. It ranges from effectiveness in digital transformation to participatory design and transparency of change, thus touching principles and structures of human life and socialization. Consequently, we dedicate the first part of the book to a system and systemic understanding of the reality we perceive and design in the course of intensifying digitalization, in particular when utilizing the full potential of upcoming generations of Cyber-Physical Systems (CPS). We start out reflecting on developments of so-called humanoid socio-technical systems that increasingly mimic human behavior including generating novel behavior and structures of digital intelligence.

We proceed with introducing relevant concepts of systems (engineering). They address the dynamic nature of system, once we zoom into socio-technical systems and need to consider interactions as substantial to humanoid socio-technical systems. In that context, we explain the nature of Complex Adaptive Systems (CAS). Since these systems consist of a variety of actors who can be in control of change and development, the understanding of the concept of System-of-Systems (SoS) helps in capturing concurrent processes under the control of multiple actors.

Since many communities and bodies organize themselves through data-driven events rather than interaction, we analyze the shift from focusing on data to behavior modeling as target of development. It not only enables a unifying perspective on systems but also features stakeholder participation, in particular through no- or low-coding elements. This shift brings us to models as baseline and frame of reference for development. Once models as digital representations, like in the case of digital twins, become the focus of development, design activities will be intertwined with engineering cyber-physical ecosystems. Thereby role and task understanding is crucial for development and operation. The latter features the emergence of digital identities that can be part of CPS, requiring digital identity management techniques, affecting predictive analytics and learning.

In the second part of the book, we discuss possible incarnations of digital selves. We start to speculate about the emergence of consciousness in artificial agents. We proceed in describing the research area of agent-based social simulation. We argue for game theory as powerful theory in modeling intra-agent behavior of large crowds. Next, we describe problems of social order and we discuss whether these problems will emerge in the course of cooperating digital selves. We show that this will not necessarily be the case, as it mainly depends on the type of implementation and on advances in computer science.

Subsequently, we present modeling of emotions based on deterministic dynamics for populations arranged in a lattice. We use non-cooperative game theory as a vehicle for social simulation. We proceed by presenting the implementation of the business transaction theory as a means for cooperating on a stochastic basis. We compare the results with the deterministic approach presented previously.

The upgrading of digital twins to digital selves is discussed in Part III of the book. Here, we present a symbiosis of Parts I and II. We show under which preconditions digital selves based on digital twins are able to establish and run digital ecosystems. The final chapter in this part is dedicated to opportunities and modes of creating reflective socio-trans-human systems, exploring mutual control and continuous development.

The epilogue is congenitally speculative in its nature. We present our thoughts on future developments of artificial life in computational substrates.

Acknowledgements

We would like to thank Springer for supporting us in organizing the content and production of the book, in particular Ralf Gerstner. We also appreciate the effort of Ines Golubovic for editing and polishing the figures.

Supported by Johannes Kepler Open Access Publishing Fund.

Contents

Chapter 1
Major Historical Landmarks in Computer Science

We consider 1930 as the beginning of a new age. This period is characterized by developments which can be expressed by exponential curves.[1] By means of digital computer systems organized in networks, information can be stored, retrieved, and distributed simultaneously to the crowd extremely speedy. This is the precondition for the quick exchange of information between different scientific disciplines and paves the way for exponential growth of ideas, inventions, innovations, and products. Furthermore, each real event, process, or thing can be expressed, modeled, or simulated in virtual environments.

In our mind, three people, namely, Kurt Gödel, Alonzo Church, and Alan Turing, can be identified as godfathers for that period. Alonzo Church (1941) invented the *lambda calculus,* which can be regarded as the basis for many programming languages far before the first computer system had been built. Alan Turing (1937), who was a scholar of Church, verbalized the somehow complicated formalized findings of Kurt Gödel (1931).Turing replaced the formal language from Gödel by a simple mechanism consisting of a *tape* divided into cells, a *head* that can read and write symbols on that tape, and a *state register* which stores the state. This mechanism models a machine on a piece of paper—later called the Turing machine. The Church-Turing thesis hypothesizes that anything that can be "computed" can be computed by some Turing machine.

In the year 1941, 10 years after Gödel's publications, Konrad Zuse invented the first programmable Turing-complete computer system Z3. He applied the first high-level programming language, which he named *Plankalkül* (Giloi, 1997). For the calculations, the Z3 computer used circuits consisting of relays and vacuum tubes.

The Boolean algebra (Boole, 1854, 2016) became the mathematical basis of digital computing. Claude Shannon (Shannon, 1938) and Victor Shestakov both showed that electronic relays and switches, called gates, can realize the expressions

---

[1]Exponential curves are not well understood by humans. In primitive times, humankind was only confronted with linear problems.

© The Author(s) 2022
F. Barachini, C. Stary, *From Digital Twins to Digital Selves and Beyond,*
https://doi.org/10.1007/978-3-030-96412-2_1

of Boolean algebra. This was the foundation of practical digital circuit design. Finally, in 1945, John von Neumann proposed a computer system architecture consisting of a memory, a control unit, and an arithmetic logic unit equipped with input and output facilities. This architecture formed the basis for the modern digital computer.

Soon afterward, the bipolar transistor based on silicon replaced vacuum tubes. The next generation of digital computers used integrated circuit chips. The invention of the MOSFET transistor enabled the practical use of MOS transistors as memory cell storage elements. From the year 1966 onward, memory capacity increased by a factor of 10 every fourth year. Following this exponential curve, we can conclude that in the year 2028, the complete storage capacity of an average human brain will be stored on a single memory chip. In order to prove this claim, one only needs to solve the following equation:[2]

$$5 \times 10^{15} = 10^{(\text{YEAR}-1966/4)}.$$

In parallel, due to rapid MOSFET scaling, MOS IC chips rapidly increased in complexity at a rate predicted by Moore's law (Moore, 1965). There is a debate among scientists about the life span of Moore's law. Due to advances in lithographic techniques, new materials, and 3D wafers, we think that Moore's observations remain valid for at least another decade. However, at the end, we will push physical limits due to tunnel effects of semiconductors. However, new technologies, such as quantum computer systems, boost computing space and power to higher limits.

In parallel to the exponentially growing hardware advances, there was a need to express algorithms in a more elegant and understandable way.[3] Therefore, new programming paradigms and programming languages were invented. During the 1950s, FORTRAN, a language for solving scientific problems, dominated. PL1 and COBOL were applied to commercial problems. During the 1970s, PASCAL and its derivatives dominated the programming scene. In addition to the imperative paradigm, functional and declarative languages, such as LISP and PROLOG, were applied. Around 1980, the object-oriented paradigm prevailed. Languages as ADA, C++, and JAVA, its pendant C#, and finally scripting languages completed the 50 years lasting development process. In the field of software engineering, software development environments and the Unified Modeling Language (UML) as a way to visualize the design of a system became common.

Tim Berners-Lee invented the World Wide Web in the year 1989. This pioneering innovation provided the capability to connect all worldwide operating computer

---

[2]The YEAR is the unknown. The left side of the equation represents the number of bits we would need to store the brain capacity. An average brain consists roughly of 10^{12} neurons. Each neuron connects to 10^3 synapses. Thus, there are in total 10^{15} synapses. One synapse can be represented in 4 bytes. Consequently, we need 4 million GBytes. Adding 1 GByte auxiliary storage yields 5×10^{15} bytes.

[3]We remember that our first computer programs were coded in hexadecimal code and later in Assembler.

systems. Nearly unlimited storage and computing power became accessible from any point in the world via cloud computing. Web pages were presented as hypertext documents using the hypertext markup language (HTML). These pages were addressed through uniform resource indicators (URI). Soon it became clear that each item in the real world might be addressable via URI's—this was the birth of the Internet of Things (IoT). The IoT plays a major role in the development of digital twins, as we will see later in this book. And for digital selves, the implementation of avatars in the second life architecture in the year 2003 was a first practical step toward the virtualization of personalities in the web.

The connectivity of the web and the availability of nearly unlimited storage result in data oceans which cannot be controlled by humans exclusively. As a consequence, the buzzword "big data" represents a synonym for algorithms analyzing these data oceans. Similar to real life, "life" in a computer system means permanent sorting and searching.

It is well known that all the programming paradigms can be converted into each other. Thus, they serve as a means for the expression of algorithms. The above-described advances in computer science form the basis for solving one fundamental question which was posed by Alan Turing (1950) in the year 1950: "Can machines think?". This question inspired numerous scientists, and the competition trying to solve the Turing test (Turing, 1950) had begun. This was the advent of artificial intelligence (AI).

Although Joseph Weizenbaum (1976) was convinced that a computer will never gain wisdom, he implemented one of the first AI programs, the ELIZA program. The rule-based program, implemented in the year 1966, used pattern matching to the written statements of patients to figure out how to reply. Although very simple, the program delivered astonishing results. Users had the feeling as if they were talking to a genuine psychotherapist. However, after a while, the program ended in loops because there were only a limited number of rules which were able to identify limited numbers of different words of the sentences entered via a console.

In the twentieth century, we implemented a version of ELIZA in our lab using the forward chaining production system language PAMELA (Barachini, 1991). The more rules we added to our knowledge base, the longer the user believed that he/she was talking to a psychotherapist. However, after a while, the game came to an end. Experts assert that today's *chat-bots* do not better perform than early rule-based programs.

Cognitive architectures like GPS (1972) by Alan Newell and Herb Simon or SOAR from John Laird et al. (1987) and Laird (2012) used production systems[4] to implement means-ends analysis so that more general problems could be solved. The reinforced learning algorithm in SOAR, which dynamically modified the weight of rules, and the chunking mechanism enabling primitive learning could solve planning problems. A Truth Maintenance System guaranteed the consistent status of the working memory. The architecture is still used as a basis for cognitive simulations.

---

[4]Production systems consist of rules, a working memory (facts), and an inference engine.

For rule-based programs applied to a very limited scope of application, the computer pioneer Barr et al. (1981) coined the word expert system. The first expert systems were applied to medical problems before being widely used in industry. Expert systems delivered promising results, while cognitive learning had its limits.

In parallel to the above-described hardware and software developments, search algorithms have been developed. They are able to prune large search spaces. In combination with heuristics and high-performance supercomputers, they are able to solve complex problems. Some computer scientists believe that playing chess or Go is a good measurement for the effectiveness of artificial intelligence. In the year 1997, IBM's deep blue computer has beaten the chess champion Kasparov. However, Go turned out to be a harder nut to crack. In this case, just speeding up the hardware and the implementation of search algorithms turned out not sufficient, since the search spaces are much more complex for Go than chess.

A new paradigm, artificial neural network processing, helped to solve the problem. Google's AlphaGo Zero program used self-training reinforcement learning. AlphaGo Zero played itself in hundreds of millions of games such that it could measure positions more intuitively and collect its knowledge in neural networks. In a way, the network masters the game without teaching the rules. The so-called deep neural network was implemented on a single machine with four tensor processing units. In May 2017, the program could beat the world champion.

Many other applications have been developed with similar kinds of neural networks since then. Among them is AlphaFold which was able to predict the 3D structure of proteins in the year 2020. There are approximately 10^{300} possible combinations for a single protein. It would therefore not be possible to solve this protein-folding problem with brute force algorithms in a reasonable time period, except for quantum computers which are in the experimental phase right now.

Similar good results were presented by IBM's Watson supercomputer in 2011. This computer was doing better than any human when playing the *jeopardy* game. Encyclopedias, dictionaries, and thesauri were stored in local memory. The machine used different computing paradigms for problem-solving, among them neural networks. Numerous commercial and medical applications are currently on its way.

According to our observations, artificial neural networks were the last cornerstone for artificial learning. They go back to Frank Rosenblatt (1958) and to Hebb's rule (Hebb, 1949). The rule explains human learning, in which simultaneous activation of cells leads to pronounced increases in synaptic strength between those cells. Much later, Teuvo Kohonen and others (Kohonen, 1982, 1984; Kohonen et al., 1991) presented forward and backward propagation algorithms for artificial neural networks. Having a closer look to the near future, autonomous cars, robots, and drones rely on this technology as well.

Whenever an article about a new AI-driven product is published, one can be confident that in most of the cases, neural networks are part of the game. However, the application of neural networks follows no cooking recipe. It is rather an experimental trial-and-error procedure until an appropriate network consisting of the right number of nodes, layers, and propagation functions is working properly.

Furthermore, engineers need to figure out how to present the data to the network and how to get the right sort of output that can be translated back into the real world.

Undoubtedly, deep learning with neural networks has brought new insights. AlphaGo Zero was able to produce strategies which were not known and applied by humans before. The same is true for radiological AI programs. They interpret pixels in different ways than humans. Therefore, we suspect that artificial intelligence deviates from human intelligence. However, if AlphaGo Zero would be confronted with a change of the board size during the game, the program would fail to deliver meaningful strategies. Thus, robustness is still a challenge for neural network applications. But there is no doubt that artificial intelligence can produce better and other results than human intelligence in limited domains.

Today's implementations combine neural networks with rule-based, object-oriented, procedural, and functional approaches. However, the mix of the paradigms and the increasing architectural complexity make software vulnerable and error-prone. Gödel has shown that the set of Gödel numbers of functions whose domains belong to a class of recursively enumerable sets is either empty or contains all recursively enumerable sets. In other words, only trivial properties of programs are algorithmically decidable. Thus, the logical correctness of a program can be guaranteed only for very simple pieces of code. Therefore, we need to implement redundancy in our architectures, like it is provided by our natural brains.

When we come back on Turing's question whether a machine can think, then the answer is ambivalent at the moment. Additional research in neuroscience is necessary. As long as neuroscience is not able to isolate the process that gives rise to consciousness in human brains, it will not be possible to create machines that have consciousness and mind, unless consciousness is a result of the thinking process. John Searle's Chinese room experiment (Searle, 1980) clearly underpins this position. On the other hand, the Chinese room experiment encourages us to simulate understanding on computers in the sense of a weak AI. There is a difference whether the computer simulates a property or whether the machine understands what the simulation is about. In the latter case, we speak about strong AI. Eventually, new findings about the human brain will result from the *human brain project* which started in 2013 and ends in 2023.

The Turing test (Turing, 1950) requires that a machine must be able to execute all human behaviors, regardless of whether they are intelligent or not. That means that a machine that can think must include behavior related to emotions, cheating, lying, etc. Mind, understanding, and consciousness are elemental properties of thinking. In the second part of this book, we discuss if and to what extent digital selves might be able to adopt such natural properties.

From a long-term perspective on human civilization, i.e., a future time period in which human civilization could continue to exist, we refer to *technological transformation trajectories*. According to Baum et al. (2019), thereby "radical technological breakthroughs put human civilization on a fundamentally different course" (p.53). Some developments might persist as status quo, broadly similar to the current state of civilization, some cause significant harm (catastrophe trajectories), and some of them may lead to expansion of the human civilization toward other parts of the cosmos (astronomical trajectories). Which actions should be pursued depends on our existing portfolio and how reflective our practice develops toward systemic responsibility. In this book, we review fundamental parts of our current portfolio and reflect on capabilities that could actively shape our future.

Part I
Digital Twins: Advent and Trans-human Development

This part lays a foundation to develop a socio-technical understanding of the nature of digital twins as representations and trans-human development objects.

Chapter 2
Background and Foundations

In this section, we provide some data on the origin of digital twins and provide insight why we consider them crucial for today when it comes to digital transformation. We also are projecting future developments in the context of humanoid socio-technical systems.

Grieves, who had raised the term in 2003, in his white paper "introduces the concept of a 'Digital Twin' as a virtual representation of what has been produced. Compare a Digital Twin to its engineering design to better understand what was produced versus what was designed, tightening the loop between design and execution" (Grieves, 2014). Given the background in industrial production, twin developers aim to increase transparency, and trigger efficiency gains, mostly in terms of faster process execution times, increased flexibility of organizing processes, and higher levels of security and safety.

As such, digital twins do not only designate a new technology but rather stand for the methodological integration of heterogeneous components, requiring protocols for exchanging even large amount of data between various technologies. Originally designed as virtual double of physical products, digital twins are increasingly becoming digital replicas of Cyber-Physical Systems (CPS). Not only the simulation of processes over the entire life cycle of a product but also envisioned architectures become core activities when CPS evolve (cf. Zhuang et al., 2017; Grieves, 2019).

A digital twin is consequently a virtual representation of a process, service, or product with parts of it in the real world. There are no predefined limits to what developers transform by digital means and represent in the virtual world: objects like planes including all kinds of operation characteristics, regions or urban infrastructures, or biological processes copying the human organism. Developers simulate functional behavior with the digital twin, in order to determine how capabilities can be changed depending on different parameters. This information helps to modify any product or service in a purely virtual manner before instantiating it for operation.

Following the original concept, essential building blocks of a digital twin are physical products or elements that are equipped with Internet-based sensors. These sensors record numerous development data and, later on, operational data in real

© The Author(s) 2022
F. Barachini, C. Stary, *From Digital Twins to Digital Selves and Beyond*,
https://doi.org/10.1007/978-3-030-96412-2_2

time, once the product is in use. A collector component records all behavior data and stores it in a centralized repository, mainly using cloud services and some digital platform for processing these data. Intelligent actors contain AI algorithms to evaluate the data. Actors could be equipped with interactive access, e.g., in product design with 3D technology or virtual reality, to visualize data and to facilitate simulating certain product properties.

In addition to the continuous optimization of the product during operation, the feedback of the usage data to the producer—often also called digital thread—enables important findings for further development and the concerned developers. This knowledge is helpful, e.g., when it comes to adapting product features better to the needs of the users and costumers. Most applications of digital twins focus on virtual models created on the basis of existing operational and design data to study the behavior of certain components at different conditions. Due to virtual prototyping, engineers and designers work close together across departmental boundaries.

2.1 Humanoid Socio-technical Systems: The Next Level

Before focusing on digital twin representations and digital identity development, we need to take a closer look on the type of systems digital twins could represent and they may become part of. Before the Internet of Things (IoT) and Industry 4.0 (I4.0) have been introduced, socio-technical systems have been composed of technical and social subsystems that are mutually related to accomplish (work) tasks within social systems, such as organizations. "Technical" thereby refers to structure and a broader sense of technicalities. "Socio-technical" refers to the interrelatedness of *social* and *technical* aspects of a system the technical and social systems are part of, e.g., an organization, an enterprise, an institution, or a sector. This context determines the relations between the technical and social systems, e.g., termed user interface between interactive technical systems and humans. It has become a central part of technology development, establishing dedicated areas such as usability engineering and user experience (UX).

The next level in socio-technical system development is targeting the merging of man and machine[1] as digital systems will cross over with material ones in many different facets. In that context, trans-humanism and singularity are crucial concepts as they allow passing on the control over further developments to artificial systems. Hence, in this section, we will start detailing some underlying assumption of trans-human developments.

We ground our system understanding on the layers shown in the figure on the architecture of digital selves in cyber-physical ecosystems. CPS are grounded on IoT- and data-related infrastructure components. They comprise micro-electronic

---

[1] See "The Merging of Man and Machine" (www.facebook.com/TranshumanismandTheSingularity).

Cyber-Physcial Systems
- Digital Twins (product prototyping – …. Shadow …. – Thread (real time simulation & operation)
- Connectivity services
- Servitization capabilities
- (Behavior) Modeling notation
- Execution support for dynamic adaptation

Infrastructure and Resources
- Micro-electronic and sensoric systems as enabler of IoT systems
- Internet-(compatible) communication protocols / stacks
- Cloud-based computing capabilities
  - Development facilities
  - Data management
  - Production capabilities

Fig. 2.1   Digital selves—architecture

and sensor systems for intelligent products, services, and production including manufacturing. They also include the provision of ICT resources, in particular for Internet-related communication and service interaction, as well as data management services (Fig. 2.1).

Digital twins as model representations of CPS can have various flavors, ranging from early-phase product development to real-time models of CPS for simulation and operation (cf. Jones et al., 2020). They require modeling and execution capabilities according to the purpose. They need to include adaptation capabilities at runtime, in order to mirror real-world behavior and influence actuators in real time. Since CPS are composed of heterogeneous components, several layers of abstractions can be required for digital self/selves development and operation. They enable communities to organize their interaction on a collective or societal basis.

Digital selves enrich digital twins with socio-cognitive algorithms. When adapting to situations, they become digital socio-cognitive counterparts of actors in the real world. As part of a socio-technical system, they operate as decision-making and support agents depending on the state of interactions. Like digital twins, they can operate at any stage of development, ranging from ideation to recycling.

This enrichment of digital twins refers to digital humanism, as it concerns all kinds of objects, including functional roles, handling of work objects and products, and judgment of events and situations. Whenever a situation requires social judgment and/or social interventions, digital selves collect and analyze relevant data to create socio-cognitive behavior using digital twin capabilities.

Like digital twins, digital selves can be used in a variety of sectors relevant for socio-technical interaction, including service, logistics, and healthcare. Digital selves can be highly adaptive, and thus accompany design and release processes, and influence the creation, operation, and function of an entire socio-technical system. As socio-cognitive models are executed to implement actor- or system-specific behavior, both the effectiveness and efficiency of these systems can be influenced by digital selves similar to social systems.

For instance, in smart healthcare systems, care-taking activities can be influenced significantly. Consider robot assistance issues. For some medical homecare users, robotic assistance may be welcome since they want to keep their individual privacy at home and prefer being supported by robot systems. Others may prefer social interaction in the course of medical caretaking at home and decide for human support. Each of them will perceive effectiveness with respect to their support of choice, even when efficiency might decrease due to required technology adaptation and/or social interventions when home health services are provided.

Trans-humanism is about evolving intelligent life beyond its currently human form and overcoming human limitations by means of science and technology. Albeit claims to guide trans-human development by life-promoting principles and values (More, 1990), advocating the improvement of human capacities through advanced technologies has triggered intense discussions about future IT and artifact developments. In particular, Bostrom (2009, 2014) has argued self-emergent artificial systems could finally control the development of intelligence and, thus, human life.

An essential driver of this development is artificial intelligence. Digital artifacts, such as robots, have increasingly become autonomous, allowing them to reproduce and evolve under their control (cf. Gonzales-Jimenez, 2018). Key is their capability of self-awareness (cf. Amir et al., 2007). According to McCarthy (2004), it comprises:

- Knowledge about one's own permanent aspects or of one's relationships to others
- Awareness of one's sensory experiences and their implications
- Awareness of one's beliefs, desires, intentions, and goals
- Knowledge about one's own knowledge or lack thereof
- Awareness of one's attitudes such as hopes, fears, regrets, and expectations
- The ability to perform mental actions such as forming or dropping an intention

Some forms of self-awareness have been considered useful for digital artifacts, in particular (cf. Amir et al., 2007):

- Reasoning about what they are able to do and what not
- Reasoning about ways to achieve new knowledge and abilities
- Representing how they arrived at its current beliefs
- Maintaining a reflective view on current beliefs and using this knowledge to revise their beliefs in light of new information
- Regarding their entire "mental" state up to the present as an object and having the ability to transcend it and think about it

The recognition of the social aspect of self-awareness (cf. Amir et al., 2007) seems to be crucial, as a digital artifact may have a system theory of itself that it can use to interact with others. Originally thought of a property to be used in multi-agent systems for dealing with errors in communication, argumentation, negotiation, etc., it could be useful for reflecting on one's own development state and articulating meaningful inputs to co-creative processes. Consequently, Amir et al. (2007) have considered several types of self-awareness:

"Explicit Self-Awareness The computer system has a full-fledged self-model that represents knowledge about itself (e. g., its autobiography, current situation, activities, abilities, goals, knowledge, intentions, knowledge about others' knowledge of its knowledge) in a form that lends itself to use by its general reasoning system and can be communicated (possibly in some language) by a general reasoning system.

Self-Monitoring The computer system monitors, evaluates, and intervenes in its internal processes, in a purposive way. This does not presuppose that the monitored information lends itself to general reasoning; in fact, there may be no general reasoning (e.g., operating systems, insects).

Self-Explanation The agent can recount and justify its actions and inferences. In itself, this does not presuppose a full-fledged self-model, or integration of knowledge needed for self-explanation into a general reasoning capability." (ibid, p.1)

Cognitive architectures such as ACT-R, SOAR, or CLARION mentioned in the second part of this book take these properties into consideration. However, social cooperation does not always need the agent's understanding. There are forms of cooperation that are evolutionary, self-organizing, and unaware. Such cooperation forms do not require joint intentions, mutual awareness, or shared plans among cooperating agents. In order to investigate intra-agent behavior of crowds regarding emotions, we therefore use more simplistic agents in our own simulations.

Nevertheless, awareness of the self and other actors can be considered as one of the important factors of trans-human (system) developments. Whether self-awareness (consciousness) can be artificially created or will automatically emerge is more deeply discussed at the beginning of the second part of this book. Apart from that, from the engineering point of view, we have to ask who is in charge. We capture this issue in part III reflecting on constructing mechanisms featuring consciousness through awareness and embodied cognition.

2.2 Singularity: Control of the Next System Generation?

Disruptive digital technologies have always been hard to convey due to unpredictable future developments based on those technologies crossing boundaries. High impact on society or on human existence as currently perceived requires to reflect on the limits of human boundaries that could disrupt the system that we live in. Singularity is such a cross-border issue. Trans-humanism as the integration of technology with our human biological computing systems considers human beings as physical bodies powered by electrical impulses controlled by the brain as computing machinery. It allows fine-tuning behavior within societal limits. These limits might not be valid for trans-humanist systems. Artificial intelligence algorithms enable through integrating the human body unlocking potential for novel community structures redefining the meaning of being and creating.

As detailed by Spindler and Stary (2019, p.1307), "singularity (more precisely, technological singularity) is a hypothetical future point in time at which technological growth becomes uncontrollable and irreversible by humans—the control and capability of further development is assigned to the digital system, resulting in unfathomable changes to human civilization. It will touch all humans, the entire human system as an entity."

In their 2013 video, Steve Aoki and Ray Kurzweil have introduced singularity by the following sequence:

- "My Name is Kay—Singularity"
- "I wanna tell you about our future, I estimate around 2025, that we will have expanded our intelligence a billionfold by merging with the Artificial Intelligence we are creating."
- "But that's such a profound expansion that we borrow this metaphor from physics and call it a singularity."
- "We'll be a hybrid of biological and non-biological intelligence," he says. "But the non-biological part of ourselves will also be part of our consciousness" (see also https://www.techbubble.info/blog/singularity-transhumanism).

Companies, such as Dangerous Things (https://dangerousthings.com/), anticipate bio-hacking as the next phase in human (*sic!*) development (likely to be support by kits such as xNTi kit https://dangerousthings.com/shop/xnti/). Evolution is characterized by self-organized "co-"evolution: "Our bodies are our own, to do what we want with. The 'socially acceptable' of tomorrow is formed by boundaries pushed today, and we're excited to be a part of it. *We hope you will be too*" (https://dangerousthings.com/evolution/). Awareness tests such as artificial consciousness tests could help in finding out whether the resulting systems or trans-humans should be deemed morally improper to further collaborate with others (cf. Bishop, 2018).

Accelerating the evolution of intelligent life beyond its currently human form by means of science and technology is grounded in a variety of disciplines focusing on the dynamic interplay between humanity and the technology developments. On the technology side, biotechnology, robotics, information technology, molecular nanotechnology, and artificial general intelligence are at the focus of interest (cf. Goldblatt, 2002). According to the trans-humanist declaration (http://humanityplus.org/philosophy/transhumanist-declaration/), trans-humanists strive for the ethical use of technologies when developing post-human actors.

They feature uploading, i.e., the process of transferring an intellect from a biological brain to a computer system through uploading, and anticipate a point in time "when the rate of technological development becomes so rapid that the progress curve becomes nearly vertical. Within a very brief time (months, days, or even just hours), the world might be transformed almost beyond recognition. This hypothetical point is referred to as the singularity. The most likely cause of a singularity would be the creation of some form of rapidly self-enhancing "greater-than-human intelligence" " (http://humanityplus.org/philosophy/) (cf. Kurzweil, 2005).

Trans-humanist protagonists envision overcoming aging and widening cognitive capabilities mainly by whole brain emulation while creating substrate-independent minds. According to the declaration, system development should be guided risk management and social processes "where people can constructively discuss what should be done, and a social order where responsible decisions can be implemented" (see declaration item 4 in Trans-humanist Declaration—http://humanityplus.org/philosophy/transhumanist-declaration/).

Post-humans could become self-informed and self-developing socio-technical system (elements) Increasing personal choices over how individuals design their live based on assistive and complementary technologies (termed "human modification and enhancement technologies" in item 9 of the Trans-humanist Declaration http://humanityplus.org/philosophy/transhumanist-declaration/) is a vision that seems to attract many people (cf. Singularity Network—https://www.facebook.com/groups/techsingularity/). Trans-humanists envision as design entity "post-humans" through continuous growth of intelligence that can be uploaded to computer systems. When laying ground to levels of consciousness that human brains cannot access so far, post-humans could either be completely synthetic artificial intelligences or composed of many smaller systems augmenting biological human capabilities, which finally cumulates in profound enrichments of human capabilities (cf. More & Vita-More 2013).

Besides applying advanced nano-, neuronal, and genetic engineering, artificial intelligence and advanced information management play a crucial role when developing intermediary forms between the human and the post-human (which are termed trans-humans). Subjects of design and later on designers themselves due to their self-replicating capability are therefore intelligent machines in an ever-increasing range of tasks and level of autonomy. Replacing increasingly human intelligence by machine intelligence is expected at some point to create machine intelligence superior to single human cognitive intelligence, characterized by effective and efficient planning, and self-emergence (cf. Bostrom, 2014).

Human-centeredness depends on the design of trans-humans Based on the current understanding of humanoid systems, socio-technical system development becomes a multi-faceted task. Humanoid systems are human-like systems, primarily known from robotics—they are also known as androids. They have artificially intelligent system components that are matched to one another based on human models. Using their spectrum of applications, it is possible to investigate which human-like behavior of an artificial system people accept to what extent in which situations and ultimately use them meaningfully. The Pepper robotic system (see http://www.SoftBankRobotics.com) is not only able to respond to questions and give instructions on how to cope with tasks but also to take action, such as showing the way. With the ability to move even small loads, thanks to lightweight arms and "four-finger hands" (cf. Rollin' Justin http://www.dlr.de), nothing stands in the way of a pervasive networking of humanoid systems with human effective areas.

Humanoid systems therefore have a mobile base that allows autonomous operation over a long range, e.g., in contrast to mobile devices that require a physical carrier system. Their core components are motion detection sensors and stereo cameras, which enable 3D reconstructions of the environment to be created even in unstructured, dynamically changing environments. In the collaborative areas of activity and technologies that are now increasingly emerging, the humanoid system, which is fundamentally independent and operates without human support, is not used in isolation. It is to be connected to other, especially IoT applications (Li et al., 2018). Humanoid systems thus become part of federated systems, i.e., they can work both autonomously and in combination with other systems.

The design of integrated humanoid systems faces the challenge of taking different development approaches into account (cf. Stary & Fuchs-Kittowski, 2020). The increasing degree of maturity of humanoid systems through activated degrees of freedom leads to the characteristic of pursuing several goals at the same time while adhering to a task hierarchy. A robot system can, e.g., serve drinks and observe the surroundings in order to avoid collisions in the surroundings, if this functionality is to be used on the service level in socio-technical environments. However, coordination with human activities on the one hand and with technologies already in use on the other hand is necessary. It depends on the area of application whether and how people configure humanoid systems and adapt them to cope with tasks. These organizational requirements may lead to additional technical functionality based on technological integration (Rosly et al., 2020).

A typical example is the home healthcare sector, which is equipped with M-IoT (Medical Internet of Things) applications (Sadoughi et al., 2020). Here, nursing staff, for example, must develop the willingness to adjust humanoid systems to patients and configure them in such a way that patient data is transferred for emergency care in order to enable effective relief. In terms of acceptance, humanoid systems become part of integrated systems.

In addition, the construction of a functional humanoid robot in the sense of a human-like artificial intelligence can lead to novel design and development tasks (cf. Andersen et al., 2019). Configuration and adaptation become a learning process (cf. Stary, 2020a) that is implemented in a humanoid system with the help of self-learning algorithms. Participation in social processes can be influenced on the basis of observations and learning psychological data (cf. Can & Seibt, 2016). This means that, in addition to the functional activities, social tasks could be addressed by humanoid systems.

2.3   Complex Adaptive Systems

When social tasks can be taken over by digital systems, the entire socio-technical system is continuously changing. Since the nature of each actor can be of different kind, either technical, social, or in between, we need to conceptualize such system according to the resulting complexity and dynamic change. Consequently, we adopt the concept of Complex Adaptive Systems considering socio-technical systems incorporating trans-humanist and cyber-physical developments.

Complex Adaptive Systems (CAS) started in the USA to oppose the European "natural science" tradition in the area of cybernetics and systems (Chan, 2001). Although CAS theory shares the subject of general properties of complex systems across traditional disciplinary boundaries (like in cybernetics and systems), it relies on computer simulations as a research tool (as pointed out by Holland, 1992, initially) and considers less integrated or "organized" systems, such as ecologies, in contrast to organisms, machines, or enterprises. Many artificial systems are characterized by apparently complex behaviors due to often nonlinear spatio-temporal interactions among a large number of component systems at different levels of organization; they have been termed Complex Adaptive Systems (CAS).

CAS are dynamic systems able to adapt in and evolve with a changing environment. It is important to realize that there is no separation between a system and its environment in the idea that a system always adapts to a changing environment. Rather, the concept to be examined is that of a system closely linked with all other related systems making up an ecosystem. Within such a context, change needs to be seen in terms of co-evolution with all other related systems, rather than as adaptation to a separate and distinct environment (Chan, 2001 p.2). CAS have several constituent properties (ibid., p.3ff—citation marked *italics*):

- *Distributed Control: There is no single centralized control mechanism that governs system behavior.* Although the interrelationships between elements of the system produce coherence, the overall behavior usually cannot be explained merely as the sum of individual parts.
- *Connectivity:* A system does not only consist of relations between its elements but also of relations with its environment. Consequently, a decision or action by one part within a system influences all other related parts.
- *Co-evolution with Co-evolution: Elements in a system can change based on their interactions with one another and with the environment. Additionally, patterns of behavior can change over time.*
- *Sensitive Dependence on Initial Conditions: CAS are sensitive due to their dependence on initial conditions. Changes in the input characteristics or rules are not correlated in a linear fashion with outcomes. Small changes can have a surprisingly profound impact on overall behavior, or vice versa, a huge upset to the system may not affect it. . . . This means the end of scientific certainty, which is a property of "simple" systems (e.g., the ones used for electric lights, motors, and electronic devices).* Consequently, socio-technical systems are fundamentally unpredictable in their behavior. *Long-term prediction and control are therefore believed to not be possible in complex systems.*
- *Emergent Order: Complexity in complex adaptive systems refers to the potential for emergent behavior in complex and unpredictable phenomena.* Once systems are not in an equilibrium, they tend to *create different structures and new patterns of relationships. . . . Complex adaptive systems function best when they combine order and chaos in an appropriate measure*—this phenomenon has been termed

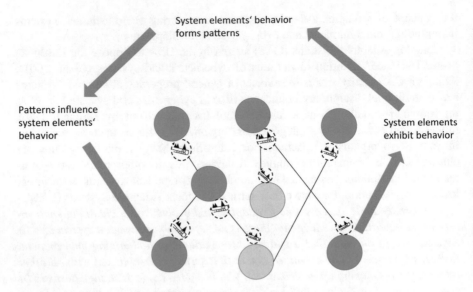

Fig. 2.2 Schema of a Complex Adaptive System and its dynamic behavior

Far from Equilibrium. CAS in their dynamics combine both order and chaos and, thus, *stability and instability, competition and cooperation,* and *order and disorder*—being termed *State of Paradox.*

In the CAS figure, a schema of a complex socio-technical system is shown at the bottom as a group of different types of elements. They are far from equilibrium, when forming interdependent, dynamic evolutionary networks that are sensitively dependent and fractionally organized (cf. Fichter et al., 2010). Taking a CAS perspective requires system thinking in terms of networked, however modular, elements acting in parallel (cf. Holland, 2006). In socio-technical settings, these elements can be individuals, technical systems, or their features. Understood as CAS, they form and use internal models to anticipate the future, basing current actions on expected outcomes. It is this attribute that distinguishes CAS from other kinds of complex systems; it is also this attribute that makes the emergent behavior of CAS intricate and difficult to understand (Holland, 1992, p.24)—see Fig. 2.2.

According to CAS theory, in CAS settings, each element sends and receives signals in parallel, as the setting is constituted by each element's interactions with other elements. Actions are triggered upon other elements' signals. In this way, each element also adapts and, thus, evolves through changes over time. Self-regulation and self-management have become crucial assets in dynamically changing socio-technical settings, in particular organizations (Allee, 2009; Firestone & McElroy, 2003). Self-organization of concerned stakeholders as system elements is considered key handling requirements for change. However, for self-organization to happen, system elements need to have access to relevant information of a situation. Since the behavior of autonomous elements cannot be predicted, a structured process is required to guide behavior management according to the understanding of system elements and their capabilities to change the situation (cf. Allee, 2009; Stary, 2014).

From the interaction of the individual system elements arises some kind of global property or pattern, something that could not have been predicted from understanding each particular element (Chan, 2001). A typical emergent phenomenon is a social media momentum stemming from the interaction of the users when deciding upon a certain behavior, such as spontaneous meetings. Global properties result from the aggregate behavior of individual elements. Although it is still an open question how to apply CAS to engineering systems with emergent behavior (cf. Holland, 2006), in the case of socio-technical system design, preprogrammed behavior is a challenging task, as humans or elements may change behavioral structures in response to external or internal stimuli. When (self)-organizing these development processes, usually they generate more complexity.

Chapter 3
Beyond Data: Unifying Behavior Modeling

In this chapter, we first motivate an integrative behavior-centered perspective on representing development knowledge, such as digital twins which serve as baseline and frame of reference for design and execution. Following the concepts presented in Stary (2020b), we introduce the Internet of Behavior as system carrier and advocate design-integrated engineering based on role and task understanding of active system elements. As distinct features, (1) functional and communication aspects are treated equally, and (2) models carry on all required information for automated execution. Thereby, the major development task is semantic modeling.

3.1 The Internet of Behaviors as System Carrier

Digital transformation increasingly binds individual activities to digital actions through various technologies which has led to the "Internet of Behaviors" (IoB) (Panetta, 2019, p.1) as follow-up to the Internet of Things (IoT) (Kidd, 2019, p.2). Consequently, behavior data direct activities of socio-technical systems in real time, encouraging or discouraging human behavior. For instance, a home healthcare support system can adapt its behavior to the situation at hand based on received sensor data and trigger specific actuator behavior based on algorithmic processing and data analytics. This trigger could lead to adjustments of human behavior, e.g., taking care of a certain order of using healthcare appliances (cf. Tan & Varghese, 2016).

Hence, the design of IoB systems based on behavior (specifications) is a moving target. As such, it is an immanent and pervasive engineering task. It requires technical and technological capabilities, when "by 2023, 40% of professional workers will orchestrate their business application experiences and capabilities like they do their music streaming services" (Panetta, 2019, p.4). Due to their cyber-physical nature—they are based on the IoT—IoB systems require a model

© The Author(s) 2022
F. Barachini, C. Stary, *From Digital Twins to Digital Selves and Beyond*,
https://doi.org/10.1007/978-3-030-96412-2_3

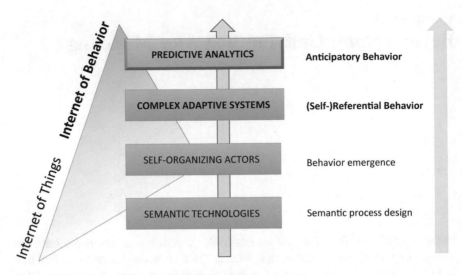

Fig. 3.1 IoB conceptualization with design intelligence

representation (termed digital twin) as baseline for continuous design-integrated engineering (cf. Lee, 2008).

Any modeling approach should enable the dynamic arrangement of networked behavior encapsulations and, thus, represent an operational framework of informed and continuous transformation. Thereby, transformation should be able to utilize IoB data for predictive analytics. Recent results indicate for specific domains the utility of algorithmic data analytics (cf. Zhiyuan et al., 2020). Featuring autonomy and digital selves, we rather need to target opportunistic IoB system behavior (cf. Guo et al., 2013), building on mutual actor awareness (cf. Gross et al., 2005) and value-based co-creation (cf. Pera et al., 2016), than unidirectional control of stakeholder or system element behavior (cf. Savitha & Akhilesh, 2020).

When aiming to identify meaningful behavior patterns, the IoB, analogous to the IoT, provides an Internet address for behavior patterns. It enables accessing systems or addressing individuals engaged with a specific behavior. Such a connection can be used in various ways and directions, for data delivery, joint processing, or taking control. Like for IoT, the power of IoB is the scale that matters. Several billions of systems and/or actors and, thus, behavior patterns populate the network and represent a unique source of collecting data and passing it on for processing, controlling, and, thus, influencing behavior through generated information.

Figure 3.1 aims to categorize the technological advancements that are characteristics of IoB developments on the left side and to develop a corresponding behavior perspective on the right side. After introducing IoT on an elementary or syntactic level, system components have been captured by semantic technologies which enabled contextual process design. Turning passive actors to active ones, and adding intelligence to system components, has led to self-organizing actors, which allowed

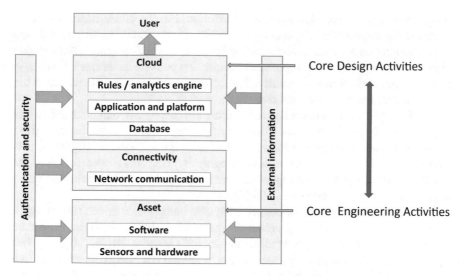

Fig. 3.2 The IoT stack (according to Kwon et al., 2016) as baseline to design-integrated engineering

the emergence of novel system behavior (Guo et al., 2011) referring to the active system elements and future developments.

Complex Adaptive Systems, as previously discussed, focus on the interdependence of behaviors. The concept raises awareness for the consequences of individual behavior on other actors or system components, as individual behavior influences the activities of other actors in the system. In this way, self-referential interaction loops develop in a specific networked setting. Understanding such a mechanism helps in the development of predictive analytics, since behavior can be anticipated based on the history of individual action and received inputs from other actors driven by those actions.

From an operational perspective, IoB systems are based on Internet-connected technologies. Thereby, the IoT architecture serves as baseline and is represented traditionally as a stack (see Fig. 3.2). IoT-based architectures facilitate interaction and data exchange between systems, their components, and users. They take into account the business perspective as well as the environment of an IoB system influencing its use and the behavioral integration of its components. The operational core consists of sensor components and the software managing them (asset part) as integrating software and hardware allows for embedded system design. Architecture components connected with the asset part are Internet components to share all kinds of collected data. They ensure connectivity of networked assets and the exchange of data. The logic to manage collected data and their transmission for processing is operated in the cloud.

Cloud computing services allow omnipresent and scalable access and distribution of system features. They comprise storing data in a database, applications and platforms to run services, rule engines to enforce (business) regulations, and analytics to generate decision-relevant information. Finally, all elements need to be related to the context of an application. It contains all relevant information for design and operation (termed external information in the stacked architecture). Another part of the stack components is composed of overarching performance-relevant topics, in particular authentication and security. Both affect the interactive and automated use of architecture components, thus running the overall system.

It is the upper part of the stacked architecture injected by external information that is crucial for design-integrated engineering (see Fig. 3.2). At some point in time, stakeholders acting in specific roles need to access the IoT system, triggering data collections or interpreting the results of analysis. They also need to know the involved component for developing and maintaining the IoT technologies, either directly or via a corresponding model (digital twin).

The ongoing proliferation of connected system components drives current application development and propagation in essential domains, such as healthcare (cf. Bhatt et al., 2017) and production industry (cf. Gilchrist, 2016). Large capabilities for intelligent system design are enabled by autonomous data collection through sensor systems, as well as the dynamic adaptation and remote control of devices through actuators. Internet-based services (cf. Bertino et al., 2016) are coupled with physical objects extending their capabilities, such as signaling the possibility of exhaustion in the case of smart shoes. The provision of such services is based on the recording of sensors and operational data, the transmission via digital networks, and the interpretation and delivery of analysis results, e.g., via apps.

When products originally designed for a specific use get enriched in scope, the design process needs to take into account further services and processes. Consider clients of a healthcare appliance including smart shoes who need to be provided with health intelligence according to their individual use of the product. Design tasks need to encounter components for monitoring and requirements articulation, leading to (dynamic) adaptation of an IoB system. It enables novel relationships between stakeholders (in particular between producers and consumers) and components, intermingling operation and development (cf. Larivière et al., 2017).

3.2   Semantic Modeling for Dynamic Evolvement

A digital twin architecture can build upon patterns of interaction as design elements for analysis and refinement to operation. Any organization can be represented as a value stream network (cf. Allee, 2003, 2009) that corresponds to a self-adapting complex system. Such a system can be modeled by identifying patterns of interactions representing relations between Behavior-encapsulating Entities (BeEs) as nodes of the network. Each BeE in a certain organizational role produces and delivers assets along acts of exchange (transactions).

Since transactions denote (organizational) task accomplishment through exchanges of goods or information, they encode the currently available network intelligence (determining the current economic success). Each BeE can be considered as IoB element in an Internet-based economy. BeEs send or extend deliverables to other BeEs. One-directional arrows represent the direction in which deliverables are moving in the course of a specific transaction. The label on the arrow denotes the deliverable. Deliverables are those entities that move from one BeE to another. A deliverable can have some physical appearance, such as a document or a tangible product, or be of digital nature, such as a message or request for information.

The concept of exchange is considered a bi-directional value stream: An exchange occurs when a transaction results in a particular deliverable coming back to the originator either directly or indirectly. It ranges from feedback on a BeE deliverable to a new request "for more of the same" or to a change of behavior. Exchanges reveal patterns typical of organizational relationships, e.g., goods and money.

In the following, we exemplify a BeE map for home- and healthcare involving a service company providing innovative instruments (methods and technologies) for customers with specific healthcare needs. The IoB system should help tracking a person's blood pressure, sleep patterns, diet, and blood sugar levels. It should alert relevant stakeholders to adverse situations and suggest behavior modifications to them toward a different outcome, such as reducing blood pressure through a different diet or reducing the dose of pills for the sake of daytime agility. Moreover, the system should provide everyday convenience, in particular alerting for timely healthcare and medical supply.

The BeE network helps scoping the design and transformation space and leverages potential changes for each BeE. Accurate service provision for wellbeing of a customer in home- and healthcare is the overall goal of the exemplified IoB system. It monitors health and living conditions to continuously improve service provision.

The first step designers need to consider in the modeling process is the set of organizational tasks, roles, or units, as well as functional technology components and systems that are considered of relevance for service provision. They represent BeEs and include the IoT devices Blood Pressure Measurement, Sleep Pattern Monitoring, Diet Handler, Medication Handler, and Personal Scheduler, as well as external medical services. Each of the identified roles or functional task represents a node in the BeE network which is partially shown in Fig. 3.3.

In the BeE network in Fig. 3.3, a specific pattern can be noticed. The Medication Handler triggers Blood Pressure Measurement, involving the Personal Scheduler to start in time. In the network without dotted transactions, Blood Pressure Management is a sink of information. Hence, in order for information not to result in "dead ends," information on blood measurement needs to be passed on explicitly to the Medication Handler and Personal Scheduler. In this way, significant knowledge can be exchanged, and further action can be designed in the case of adverse conditions.

At this stage of design, exchange relations can be added, as indicated by the dotted transactions in Fig. 3.3. In the simple example, Blood Pressure Measurement should be in exchange relations to the Medication Handler and Personal Scheduler, as the medication could be adapted and optimized according to the time and current condition of the client. It needs to be noted that this is a semantically grounded

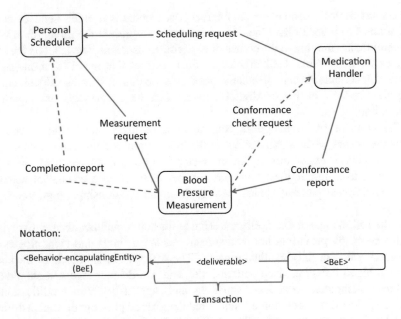

Fig. 3.3 Part of a BeE network scoping and structuring the transformation space

supplement requiring systemic domain knowledge and human intervention, in contrast to syntactically checking whether each BeE interacts with all others in the network.

Refining design representations, such as the BeE network, toward digital models of IoB systems serves as the baseline for engineering. Recognizing the behavior of networked actors or system components requires not only to capture the duality of activities, namely, in terms of functional (or technical) and communication actions, but also their autonomous while synchronized acting as networked nodes. Hence, we propose to utilize the choreographic representation and engineering scheme stemming from the Subject-Oriented Business Process Management (Fleischmann et al., 2012a, 2015). It enables embodying BeE network and refining the involved (socio-technical) components, thereby generating behavior-centered digital twins.

For design-integrated engineering, subject-oriented models can be refined until being executable, in order to provide operational feedback to designers. These modeling and execution capabilities view systems as sets of interacting subjects. Subjects are defined as behavior encapsulation. They address tasks, machine operations, organizational units, or roles people have in a specific context, such as business operation, and correspond to the Behavior-encapsulating Entities (BeEs) defined for analyzing value streams as described above. From an operational perspective, subjects operate in parallel. Thereby, they exchange messages asynchronously or synchronously. Consequently, the transactions forming value streams can be interpreted as transmissions of messages between subjects.

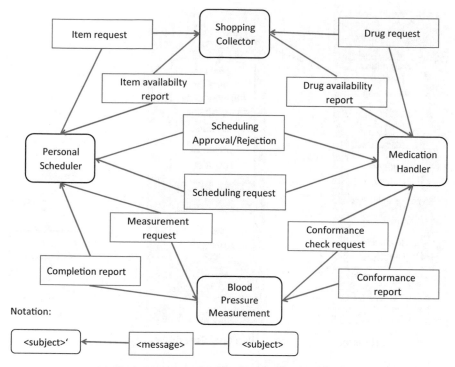

Fig. 3.4 Sample Subject Interaction Diagram representing a home healthcare appliance

IoB systems specified in terms of subject-oriented models operate as autonomous, concurrent behavior entities representing distributed (IoB) elements. Each entity (subject) is capable to performing (local) actions that do not involve interacting with other subjects, e.g., calculating a threshold value of blood pressure for a measurement device in medical care. Subjects also perform communicative actions that concern transmission of messages to other subjects, namely, sending and receiving messages.

Subjects as behavior encapsulations are specified in adjacent diagram types: Subject Interaction Diagrams (SIDs) and Subject Behavior Diagrams (SBDs). They address different levels of behavior abstraction: SIDs a more abstract one, denoting behavior entities and an accumulated view on message transmissions, and SBDs refining the behavior of each subject of a SID by revealing the sequence of sending and receiving messages as well as its local actions (i.e., functional behavior).

SIDs provide an integrated view of an IoB system, comprising the subjects involved and the messages they exchange. A part of the SID of the already introduced home- and healthcare support system is shown in Fig. 3.4. According to the SID BeEs, it comprises several subjects involved in IoT communication. In the figure, the messages to be exchanged between the subjects are represented along the links between the subjects as rectangles, already including the supplemented ones from the value stream analysis:

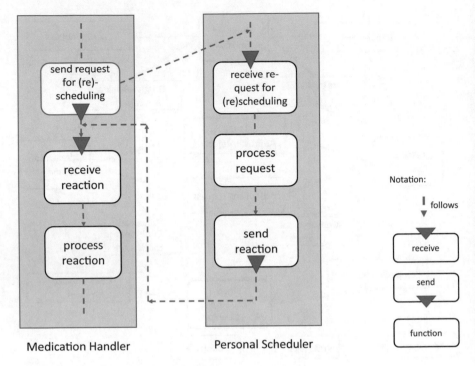

Fig. 3.5 Sample Subject Behavior Diagrams and message exchange upon request

- The Personal Scheduler (subject) coordinates all activities wherever a client is located (traditionally available on a mobile device).
- The Medication Handler takes care of providing the correct medication at any time and location.
- The Blood Pressure Measurement subject enables sensing the blood pressure of the client.
- The Shopping Collector contains all items to be purchased to ensure continuous quality in home healthcare.

The client user handles the measurement device and needs to know when to activate it and whether further measurements need to be taken. The Shopping Collector receives requests from both, the Medication Handler when drugs are required from the pharmacy, physician, or hospital and the Personal Scheduler, in case further medicine for the client user is required.

State transitions are represented as arrows, with labels indicating the outcome of the preceding state. The part shown in Fig. 3.5 represents a scheduling request to the Personal Scheduler subject sent by the Medication Handler subject, in order to demonstrate the choreographic synchronization of behavior abstractions (cf. Wen et al., 2017). The figure reveals the parallel operating nature of the two subjects involved in the interaction. Once the need for (re)scheduling—modeled as send activity—is recognized by the Medication Handler, a corresponding message is delivered to the Personal Scheduler. When the Personal Scheduler has received that message, the request can be processed, either recognizing a conflict or fixing an entry into the schedule. In both cases, the result is delivered by "send reaction" to the Medication Scheduler. The subject that has initiated the interaction can now process the results, i.e., the Medication Handler processes the reaction of the Personal Scheduler (modeled by the function of the respective SBD).

Each subject has a so-called input pool as a mailbox for receiving messages (including transmitted data through messaging that are termed business objects). Messages sent to a subject are kept in that input pool together with their name, a time stamp of their arrival, the data they transport, and the name of the subject they come from. The modeling designer can define how many messages of which type and/or from which sender can be deposited. The modeler can also define a reaction, in case messaging restrictions are violated, e.g., to delete arriving messages or to replace older messages in the input pool. Hence, the type of synchronization through messaging can be specified individually.

Internal functions of subjects process (the transmitted) data. In our example, the subject Blood Pressure Measurement has a counter for each application. An internal maintenance function increases the counter by one when the client user activates the device. The function can either end with the result "sufficient energy" or "change battery."

Once a Subject Behavior Diagram, e.g., for the Blood Pressure Measurement subject, is instantiated, it has to be decided (1) whether a human or a digital device (organizational implementation) and (2) which actual device is assigned to the subject, acting as technical subject carrier (technological implementation). Validation of SBDs is sufficient for interactive process experience and testing process completion. Besides academic engines, e.g., UeberFlow (Krenn & Stary, 2016) or SiSi (Elstermann & Ovtcharova, 2018), commercial solutions, such as Metasonic (www.metasonic.de), Compunity (www.compunity.eu), or actnconnect (www. actnconnect.de), can be used. Since neither the input pool nor the business objects are part of the modeling notation, it depends on the environment and runtime engine used for development, at which point in time, and in which form data structures and business logic determining the communication on the subject instance level can be specified for pragmatic process support.

Dynamic adaptation is based on a trigger, such as a result from performing a function or a sensor signal, which requires special behavior specification. It can be handled according to S-BPM's concept of event processing, thus allowing to capture variants of organizational behavior at design time (cf. Fleischmann et al., 2012b). The trigger to dynamic adaptation independent to its implementation can carry some data as payload. For instance, with the trigger "blood pressure above threshold," some information can be tagged to the physical device. Like an event, a data object representing a trigger can carry three types of information: header, payload, and plain content. The header consists of meta-information about the trigger like name, arrival time, priorities, etc. The payload contains specific information about the triggering event. Finally, a trigger can also contain free format content.

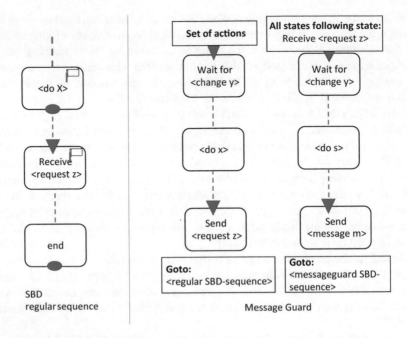

Fig. 3.6 Dynamic behavior adaptation using Message Guards

With respect to operation and model execution, triggers are messages. Messages of a S-BPM model represent event types. Once a process instance is created and messages are sent, these messages become events. If messages are sent and kept in the input pool, they get a time stamp documenting their arrival time. Instantaneous events can be handled by Message Guards. They are modeling constructs to represent behavior variants including the conditions when which variant is relevant and should be executed (see Fig. 3.6).

For instance, the message "call emergency service" from the subject Blood Pressure Measurement can arrive at any time when delivering data from measurement. This message is handled by a Message Guard. In that Message Guard, the reaction of an instantaneous message is specified, e.g., the emergency service is called by the Personal Scheduler subject, since reaching a certain threshold of the blood pressure indicates the need for medical expert intervention for the concerned client.

Message Guards as shown in Fig. 3.6 allow handling adaptive behavior at design time. The specification shows how critical cases are handled at runtime (i.e., once the subject has been instantiated), either by humans or technological systems. The general pattern reveals that jumping from routine behavior (left side) to non-routine behavior is based on flagging functions serving as triggers and (re-) entry points. In the addressed home healthcare example, the Message Guard can be applied when a threshold of Blood Pressure has been reached. Once the flag is raised at runtime, either substitutive procedures, returning to the regular SBD sequence (see

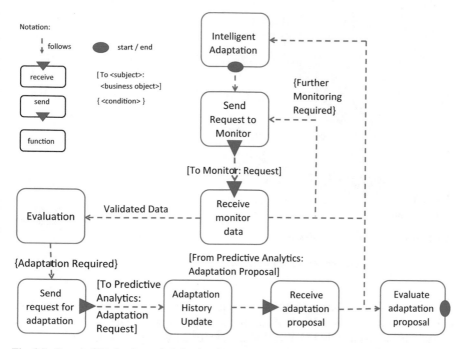

Fig. 3.7 Sample SID for dynamic behavior adaptation using Predictive Analytics

left side of Message Guard), or complementary behavior, leaving the originally executed SBD (see right side of Message Guard in Fig. 3.6), is triggered.

Message Guards can be flagged in a process in various behavior states of subjects. The receipt of certain messages, e.g., to abort the process, always results in the same processing pattern. Hence, this pattern should be modeled for each state in which it is relevant. The design decision that has to be taken concerns the way how the adaptation occurs, either extending an existing behavior or replacing it from a certain state on.

In the home healthcare example, returning to the original sequence (regular SBD sequence) is given when the called emergency service in the case of high blood pressure does not require any further intervention of medical experts. Replacement of the regular procedure, however, is required, in case the Medication Handler subject and, as a follow-up, the Personal Scheduler subject (referring to the time of medication) have to be modified.

Once the organizational transformation includes predictive analytics, its integration needs to be structured according to its context (cf. Sodero et al., 2019). Figure 3.7 shows an organizational approach for embodying predictive analytics. The developed pattern is based on a Monitor subject that is triggered by a function in an idle loop observing an IoB system. The monitored data needs to be evaluated to identify the need of adaptation. For algorithmic decision-making, a (business) rule base could be of benefit.

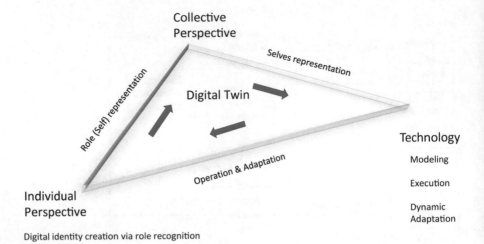

Fig. 3.8 Digital twins as behavior representations

Recognizing the need of adaptation requires business intelligence stemming from a Predictive Analytics subject. According to the behavior data available and the calculation model to either predict the behavior of the acting or the behavior of other interacting subjects, a proposal is generated. In order to avoid re-iterating certain behavior patterns, the adaptation request is stored, together with the newly generated proposal. The latter will be evaluated for effectiveness and efficiency.

With respect to our home healthcare case, organizing the setting could be challenged on whether medical experts need to be contacted once the blood pressure is higher than a specific threshold by predicting that an additional data analysis (e.g., diet patterns) could help avoiding triggering emergency services. Implementing such a proposal requires extending the SID with a Diet Handler subject that can deliver timely data on the diet behavior of the client. It would need to interact with all other subjects, as its functional behavior to provide the requested data leads to novel patterns of interaction.

Each enrichment can be iteratively tested along design-evaluation iterations of various granularities. Consequently, whenever the transformation space is to be enhanced with choreographic intelligence, the resulting additional requirements for a solution can be exemplarily met by (re-)designing the artifact, demonstrating and evaluating the envisioned enhancement. In this way, even variants of system intelligence can be explored and checked for viability.

Digital twins based on executable models enable threads for simulation and operation of IoT-based systems and have gained recent research attention (cf. Pang et al., 2021; Stary, 2021), also driven by Industry 4.0 needs for real-time operation, integration capabilities, and high-fidelity representation. From the three perspectives introduced in the beginning for structured understanding, a development cycle can be concluded for digital twins (see Fig. 3.8). Individual role understanding leads to model and structure collectives that are operated by technologies and can be adapted to further needs, including socio-emotional behavior finally leading to digital selves.

The notation has not only the character of a modeling language but rather serves as means of orientation for various stakeholder groups, including users and developers. Hence, a certain representation scheme, such as subject-oriented modeling, has specific roles for digital twin development:

- It helps in organizing a collective by means of structure elements.
- It supports the design of ecosystems as a boundary object mediating between various stakeholder groups, including domain experts, developers, and users.
- It facilitates implementation through detailing structures and processes for operation.

In the role of a user of digital twins, stakeholders make use of a systems or product and experience the functionality through its features. By putting the functionality into individual context of use, it is not only shaped by individual situations of use but also shapes its future use and actions of the user. By utilizing the features of the product, not only cognitive factors but also affective, emotional, and social aspects play a crucial role and build up experiential knowledge.

In the role as developers, humans utilize organizational and technological capabilities to refine and engineer functional requirements referring to user demands and support needs. Technology-driven activities include performance-related items aside from usability and user experience. In the case of cyber-physical settings, decisions on heterogeneity and connectivity play an important role from an engineering perspective. They influence adaptability during runtime to a large extent.

Part II
Social Behavior of Artificial Agents

Following the conceptual understanding of digital twins and how they could be engineered according to cognitive and organizational structures, this part lays a foundation for understanding social behavior and its modeling. We discuss various socio-emotional approaches before sketching behavior-relevant models and their simulation capabilities. In particular, we show how emotions can substantially influence the collective behavior of artificial actors.

Chapter 4
Background and Motivation

In Part I, we discussed mainly engineering issues and solutions to socio-technical systems dealing with digital twins. We defined properties which are mandatory for digital selves represented by digital twin elements or agents. We argued that properties of agents need to be enhanced with behavior related to emotions. In this section, we present game theoretical approaches to emotion modeling. Since we investigate the collective behavior of crowds, we are interested in inter-agent relationships. Emulations of consciousness are out of scope of our investigations. Since a very specific property is simulated, it is feasible that the agents and their learning functions are more simplistic.

When writing these sentences in the year 2021, we just realized that in Graz (Austrian city), a research team managed to implement a full-fledged digital heart (see http://healthcare-in-Europe.com). This artificial heart is produced out of numerous MR and EKG data and other heart investigations. Imaging algorithms construct the heart based on all these very special investigations conducted on a patient. To simulate the heartbeat, many million variables need to be calculated in real time on a super computer. AI algorithms support the calculation of the electrical wave propagations for the simulated heartbeat.

The development team asserts that the simulated heartbeat is already close to the real heartbeat of a patient. They plan to apply more AI methods so that the heartbeat runs completely in synchrony with a patient. However, the authors did not publish yet any details about the AI methods applied. We suspect that neural networks are playing a major role in learning the heartbeats. The authors claim that within 2 years, it will be possible to treat a digital heart with artificial drugs. As a consequence, it will be possible to decide which of the several drug treatments turns out to be adequate after the implantation of a cardiac pacemaker.

This is indeed a revolution in medicine and in computer science. We ask ourselves if this is possible for the heart, why not for the human brain. And if we try to model the human brain on a digital machine, is this copy identical to the real brain? Is the digital twin capable of thinking? Does it have its own personality? Will

© The Author(s) 2022
F. Barachini, C. Stary, *From Digital Twins to Digital Selves and Beyond*,
https://doi.org/10.1007/978-3-030-96412-2_4

consciousness emerge as a consequence of thinking or is it rather an illusion? Is there even a free will?

We can only speculate about the answers to these questions since to our best knowledge, there is no commonly approved theory of mind existing yet. However, neuroscientists are making progress in brain research. We know today that the human brain is organized in a hierarchy. Dehaene and Changeux (2011) investigated the transition from unconscious to conscious perception. They showed that according to Baar's theory (Baar, 1988), this what we experience as perception is caused through the intensification of a single information entity which is then simultaneously transferred in different parts of the cerebral cortex. By applying functional MRT and EEGs, Dehaene discovered that the transformation process triggers certain synchronized rhythms of electrical activities. Presumably, these rhythms stimulate consciousness by activating a network of pyramid neurons located in the parietal and frontal brain regions which in turn cause top-down amplification. Dehaene discovered this phenomenon while experimenting with the picture recognition process of the human brain. Similar results were obtained during experiments with the tactile sense. It can be concluded that there is a threshold between unconscious data processing and conscious data processing.

Libet (2005) also discovered that a lot of the brain activities are not conscious at all. Based on his experiments, Libet concluded that our awareness of decision-making appears to be an illusion. According to Libet, there is also no free will.

Neurobiologists suspect that changes in the hormone level not only influence feelings and spirit but also the function of the brain. Obviously, parts of the hippocampus are changing its structure and its activity during specific hormone cycles of women. The implications of these observations are not yet well understood.

Eric Kandel (2018) presents a good overview how art is interpreted by the human brain. As computer scientists, we also find Kurzweil's elaborations (Kurzweil, 2012) about the creation of a mind promising.

In order to mimic the hierarchies of the human brain, Kurzweil proposes the implementation of hidden Markov models in combination with self-learning neural networks. He is convinced that before 2050 or even earlier, we will bring the full power of human brains on silicon substrates. In contrast to our brain, artificial neural networks will learn much quicker. This is due to the fact that during the learning process, the human brain takes time to grow new dendrites. Technologically, we are able to construct artificial neural networks consisting of more neurons and synapses than the human brain. Humans cannot expand the size of brains on biological level in a short time. But thanks to exponential technological growth, we might do it with artificial neural networks in the farer future. Kurzweil has certainly a point there.

The question above, whether consciousness is an emergent property, is still not solved. It might be emergent if we follow Deahene's observations. If we consider, e.g., hormones as important influencers, then we need to understand also their behavior. As soon as we understand their influence, we can design an algorithm which can be executed as a Turing machine on a real computer. If consciousness follows an algorithm somewhere in the neo-cortex—which has not yet been proven—then we can simulate also consciousness. If consciousness emerges automatically as a result of low-level perception processes, then we will see the results of virtual humans (humanoids) or virtual agents in a few decades.

We believe that Kurzweil's ideas described more deeply in Part I will bring substantial progress in the sense that a good percentage of a brain will be simulated in silicon substrate in a few decades. And it could easily be that this brain might outperform average brains in certain general tasks due to quicker expansion options. Much more artificial neurons and synaptic connections yield more brain capacity, and the presentation of several millions of examples in a short time period yields faster learning cycles. However, as long as the influence of electro-chemical processes on synaptic circuits is not fully understood, the silicon brain might probably lack certain high-level functionalities. When writing these lines, robots can simulate the behavior of approximately 4-year-old children.

Since a digital self is a simulation, digital selves are different from one self, similar as human twins differ from each other. There is a difference whether the computer simulates a property or whether the machine understands what the simulation is about. It is therefore an illusion to think that we can replicate ourselves and thus exist for eternity.

Artificial neural networks are producing other and sometimes better strategies for problem-solving, as explained when addressing the AlphaGo Zero program. Consequently, there is an artificial intelligence which deviates from human intelligence. Thus, a digital self might develop its own artificial personality, at best.

Subsequently, we talk about agents when addressing the modeling and simulation of digital selves or artificial humans (humanoids). The term agent is a more pragmatic one and widely used in the simulation community. The simulation community mostly uses either a deductive approach or an inductive approach (Axelrod, 1997). In order to find consequences of assumptions, deduction is used. If we want to find patterns in data, we use an inductive approach.

In the deductive approach, social phenomena are represented either by rules or differential equations. However, as Troitzsch (Troitzsch & Gilbert, 2005) and Gilbert explain, differential equations are often difficult to study. There is sometimes no set of equations that can be solved to predict the system characteristics, and even more, group behavior and especially individual cognition (Conte & Castelfranchi, 1995) are hard to model. There is some parallelism when applying artificial neural networks, where engineering experience decides about the success of an implementation. We predict that simulations with differential equations and linear equations will play a major role in the future once the quantum computer is in place. A quantum computer has three main application areas: cryptography, searching and learning, and simulation. In the latter case especially linear equation, systems are very well suited for quantum computers because one basic gate operator, such as the *hadamard* operator, uses matrix operations.

The inductive approach is more often used by social scientists. This approach deduces generalizations from observations. The application of agent-based simulations observing emergent patterns belongs to this approach. However, Sun (2006) criticizes that most of these agents are often too simple. They comprise only few entities and lack cognitive modeling. As a consequence, such simplistic agent models do have problems to emulate human cognition. In his book about cognition and multi-agent interactions, Sun argues that cognitive architectures, such as ACT-R, SOAR, or CLARION, might be better vehicles to simulate human behavior because not only inter-agent processes but also complex intra-agent processes are considered.

The CLARION architecture (Sun, 2003) seems to be especially powerful since it combines built-in meta-cognitive constructs, action-centered and non-action-centered representations, top-down rule-based learning, and bottom-up reinforcement learning based on the Q-learning algorithm (Watkins, 1989). Sun argues that four distinct levels need to be orchestrated in a cognitive architecture: the sociological level, the psychological level, the componential level, and the physiological level. The sociological level simulates conventionally the relations in between agents. The psychological level covers individual experience, beliefs, concepts, and skills. The componential level looks at the different components of an agent and its different implementation methods. The physiological level focuses on elementary biological and physiological components of a system which need to be reverse engineered.

Another inductive approach is the application of game theory to social science. Game theory turns out to be a powerful theory when simulating social processes of large crowds. The theory of games was first formalized by Neumann and Morgenstern (1953). Human economic behavior has been studied since then extensively by using this theory.

Like with humans, communication and collaboration in between artificial agents are important steps further for the development of an artificial social intelligence. It was always argued by scientists that a machine that can think must include emotional aspects, because a vast amount of social behaviors in humans are emotionally driven (Castelfranci, 2000). If we want to construct human-like machines, then we need to integrate emotional aspects into our models. In the sequel, we show how emotions can be modeled, and we show how emotions can substantially influence the collective behavior of artificial agents.

We model the impact of rewards on the emotion *jealousy* from two very different perspectives, and we consider the population dynamics and equilibrium which can arise. First, we apply a deterministic model and then a stochastic one. Our modeling technique starts with one simple game theoretical approach. We use this approach because the social behavior which is expressed on inter-agent relations can be elegantly simulated.

In Chap. 5, we apply a more complex approach. In the sense of Sun, we model the sociological level and the componential level. We show how different kinds of emotions can be combined. However, cognitive details of intra-agent processes as explained by Carley and Newell (1994) are not considered in our models.

As explained, cognitive architectures do have their merits. Nevertheless, the problem how emotions can bypass decisions by activating some impulses (Löwenstein, 1996) is not well understood. It has also not yet been solved by these architectures. Furthermore, it is not known how emotions can influence cognitive decision processes (Löwenstein, 1996). As long as these problems are not solved by neuroscientists, we have to be careful in applying them in cognitive architectures.

But serendipity effects can always occur. We are convinced that in the very long run, such multi-paradigm architectures might become obsolete because symbolic computing such as rule-based learning, planning, storing of data, and even remembering are implicitly performed by a single morphological architecture—the human brain.

For instance, *remembering* can be modeled with stochastic dynamic artificial networks known as bi-directional associative memories by using the Hopfield model (Hopfield, 1982, 1984; Hopfield & Tank, 1985, 1986). Consider the visual cortex as a region in the brain where many neurons together decode visual sensory perceptions as an example. There, the three-dimensional information is mapped on a two-dimensional field of neurons in the human cortex (Glickstein, 1988). Such structures can be modeled with a self-organizing Kohonen network (Kohonen, 1982, 1984; Kohonen et al., 1991). Similarly, different other brain regions demand other neural network topologies and corresponding learning algorithms. But all of them are based upon a morphologic structure consisting of neurons, dendrites, and synapses. Therefore, intra-agent processes are likely to be modeled exclusively with artificial neural networks in the long run.

But we also have to keep in mind that the general learning problem of neural networks is intractable. The size of the learning problem in artificial neural networks depends on the number of unknown variables which are represented through weights of the edges between the neurons and the synapses. These unknowns need to be calculated. A network with 100 unknown weights represents a bigger learning problem than a network with 10 weights. A polynomial runtime of the learning algorithms dependent on the number of unknown variables would be desirable. However, until recently, there is no algorithm known that such an algorithm exists. Therefore, we still face some NP-complete problems.

The human brain consists of approximately 10^{15} synapses. Thankfully, not all neurons are mutually connected. As explained, there is a hierarchical structure in the brain's morphology. Certain regions are responsible for certain tasks. But there are still vast amounts of local connections left so that we can run into these NP-complete problems during the learning phase which is continuously performed in real time. However, we are convinced that in the very near future, quantum computers will overcome some of these problems.

Time has not yet come to model all cognitive processes with neural networks. We still lack a common theory of mind—we don't know how hormones are influencing synaptic thresholds—and we still need more powerful computers so that the parallel thinking process can be modeled in real time. It is not the size of the shrinking memories which represents the problem, but the runtime required to perform respective actions. Nevertheless, for the simulation of cognitive intra-agent processes, such as values or beliefs, one could start with a simple neural network and extend this network similar to a bootstrapping process until a powerful artificial brain is formed.

Social cooperation does not always need the agent's understanding as defined by Amir et al. (2007) and explained in Part I of this book. There are forms of cooperation that are evolutionary self-organizing and unaware. Such cooperation forms do not require joint intentions, mutual awareness, or shared plans among cooperating agents (Macy, 1998).

We are interested how emotions are changing the collective behavior of a group and vice versa. We don't plan to emulate human cognition as long as there is no widely recognized unified theory available. Therefore, our agents are more simplistic. When we equip agents with emotion-like behaviors, they will show similar behaviors as humans. Jealousy might be influenced by rewards. Therefore, we examine the impact of rewards to jealousy. We show that different modeling techniques and different algorithms can yield deviating results for one and the same property. The phrase "what you see, you will be" is valid for humans as well as for agents.

It is therefore in our hands whether the future of artificial humans (humanoids) will be bright or rather dark in Leviathan's sense. This reminds us on the old Indian allegory where a father tells his child that there are two animals fighting in his breast. One animal is full of negative emotions like hate and jealousy; the other animal is obliging and friendly. On the child's question which one is going to win the fight, the father's answer is "the animal which you feed is going to win."

Consequently, the data and the world which we are presenting to the programs and the algorithms we are implementing do have impact to the behavior of artificial agents. The same relation holds for humans—experience, education, and epigenetic on the one hand and inherited properties on the other shape our behavior. But in contrast to computer programs revealing rational behavior, humans often act in an irrational way, since they are driven by emotions as well. We discuss this phenomenon in Chap. 6 of Part II.

Chapter 5
Simulation Methods and Game Theory

In this chapter, we present the areas constituting agent-based social simulation, and we give a short overview on game theory. The given explanations should sufficiently contribute to the understanding of the next chapters. We argue why game theory is such a powerful theory in modeling intra-agent behavior of large crowds.

In general, a simulation is an imitation of the operation of an object or process over discrete or continuous time. A prerequisite for a simulation is the availability of a model. As explained in part I of this book, the model represents the characteristics of the object or process. The simulation reflects the evolution of this model over time. A computer system developer first creates a design of the model, then he/she implements the model itself, and finally he/she executes the simulation of the model on a computer system.

Differential equations are used in a continuous simulation which is based on continuous time. Discrete event simulation is used for systems whose states change their values only at discrete times. Stochastic simulation is a simulation where some variables of the object are subject to random variations.

Social simulation studies issues in social sciences. Beside inductive and deductive approaches, Robert Axelrod (1986) paved the way for a third approach. In his experiments, he generated data that can be analyzed inductively, stemming from specified rules rather than from direct measurements of the real world. Since then, agent-based simulation became popular among social scientists. In contrast to system-level simulation where the scientist looks at the situation as a whole, agent-based simulation consists of modeling societies with artificial agents. The properties of these agents might be simple or more complex. Based on the behaviors of these agents, collective behavior patterns can emerge. In this way, researchers can study social phenomena which are derived from individual behavior.

Agent-based simulation is appropriate for the study of processes which lack central coordination. However, looking critically to this approach, we admit that too simple agents might predict human behavior in an oversimplified manner. Complex agents on the other hand might introduce intractable organizational and computational issues.

© The Author(s) 2022
F. Barachini, C. Stary, *From Digital Twins to Digital Selves and Beyond*,
https://doi.org/10.1007/978-3-030-96412-2_5

Fig. 5.1 General payoff
matrix

		Player P1	
		S1	**S2**
	S1	a, a	b, c
Player P2			
	S2	c, b	d, d

Troitzsch and Gilbert (2005) present an accurate introduction of tools and simulation methods for social scientists. In our practical examples, we use a spatial approach, where all agents are arranged in a lattice, and we present a stochastic approach where agents meet randomly. Both examples are discrete event simulations using an evolutionary game theoretical approach.

Game theory distinguishes between two approaches. In cooperative game theory, the players are allowed to communicate and to bargain. In non-cooperative game theory, direct communication between players is not possible. Subsequently, we discuss non-cooperative game theoretical approaches because our simulations are based on non-cooperative game theory.

To represent games in a computer system, researchers use either the extensive form or the normal form. The extensive form can be used to formalize games with a time sequencing of moves. The games are formalized by trees. In our examples, we use the normal form for game representation.

The normal form is represented by a payoff matrix which shows the players, strategies, and payoffs (see Fig. 5.1).

Player P1 chooses the column, and player P2 chooses the row. Each player can choose between two strategies (S1, S2). The payoffs are provided in the interior. If player P1 chooses the strategy S1 and player P2 chooses S1, then both players get the payoff a. If player P1 chooses strategy S2 and P2 chooses S1, then player P1 gets the payoff c, and player P2 gets b. It is exactly the other way around if player P1 chooses S1 and player P2 chooses S2. If both players choose S2, then both get d.

A very special situation occurs when the following relations are true: $c > a > d > b$ and $a > (c + b)/2$. In this case, the game is a prisoner's dilemma game. This game captures the essential problem of cooperation. Axelrod (1984) examined under which conditions in a world of egoists without central authority cooperation could emerge.

In the prisoner's dilemma game, two people are suspected of having committed a crime. Since the two suspects are isolated, they cannot talk to each other. The attorney offers them a deal. If one suspect confesses while the other does not, then the first will go free, and the second will be sentenced to 5 years. If both confess, they will both receive 4 years. If neither confesses, then both will be sentenced to 2 years only. The corresponding payoff matrix is given in Fig. 5.2. Negative numbers represent the loss of free years.

This matrix game can be repeated several times. From the matrix, we can derive that cooperation is vulnerable to exploitation by defectors. If both persons analyze the game in a rational way, then both will defect (confess). Since in case one

	not confess	confess
not confess	-2, -2	-5, 0
confess	0, -5	-4, -4

Fig. 5.2 A prisoner's dilemma matrix

confesses and the other one does not, then the one who confesses goes free, and the other one will go to prison for 5 years. When the other player thinks the same, they both will lose 4 years. And this is the dilemma. Rational players who act in order to maximize their payoff defect, but it would be better to cooperate and therefore not to confess. In this case, both would end up losing 2 years only which is much better than losing 4 or 5 years.

In game theory, a player always tries to maximize his/her fitness. In the empirical experiments with humans, it can be shown that humans do not always behave in a rational way. In the repeated prisoner's dilemma game, humans often try to cooperate, and only when they learn that their strategy fails, they switch to defection. Several social dilemmas are of prisoner's dilemma type as explained in the next chapter.

In his outstanding book, John Maynard-Smith (1982) introduced game theory to evolutionary biology and population dynamics. In evolutionary game theory, individuals have fixed strategies, and they interact randomly with each other. Payoffs are interpreted as fitness, and all the payoffs of these interactions are added up. Successful games are interpreted as reproductive success similar to natural selection. Strategies with better payoffs reproduce faster, and strategies with lesser payoffs are outcompeted.

Thus, with a simple shift in the interpretation of the payoffs, we can simulate the spread of a virus or animal behaviors in a limited habitat. This makes game theory so powerful. For evolutionary game dynamics, differential equations such as the replicator equation (Hofbauer & Sigmund, 2003) are applied. The equation describes frequency-dependent selection among different strategies in infinitely large populations. Game theory in combination with this differential equation makes simulation highly effective. We are not going deeper into this subject because this is not necessary for the understanding of the rest of the book.

Two discoveries are central in game theory. In a finite population with a finite number of strategies, there is always a Nash equilibrium (Nash, 1950). Maynard-Smith's discovery of evolutionarily stable strategies (ESS) is a similar concept to the Nash equilibrium.

The Nash equilibrium indicates that if all players play a strategy that happens to be Nash, then neither person can deviate from her strategy and increase her payoff. In the payoff matrix of Fig. 5.1, S1 is a Nash equilibrium if $a > c$ or $a = c$, and S2 is a Nash equilibrium if $d > b$ or $d = b$.

An ESS is a strategy with the following property: if all members of a population adopt the ESS, then no other mutant strategy is able to invade the population. A mutant is a so-called invader. A domestic strategy is an ESS, if one of the following two conditions applies:

1. When someone from the domestic population meets someone from the domestic population, they get in average more payoff than a mutant strategy will get when meeting a domestic member. In this case, mutants cannot spread.
2. If the domestic strategies get the same payoffs among each other as a mutant strategy against a domestic strategy, then the following must apply: the domestic strategy gets more when meeting a mutant strategy than the mutant strategies get among each other.

To put it more formally: if P is the payoff and I indicates a domestic strategy and J a mutant strategy, then I is ESS if one of the two conditions applies:

1. Either $P(I, I) > P(J, I)$ or
2. $P(I, I) = P(J, I)$ and $P(I, J) > P(J,J)$

(1) and (2) must be fulfilled for each possible alternative J. If this is the case, then a domestic strategy cannot be invaded by a single invader.

If a strategy is an ESS, then it is also a Nash equilibrium. Maynard-Smith and Nash made their discoveries independent from each other.

In Fig. 5.2, we presented only perfect strategies—that means that the players can select either one or the other strategy. In reality, in repeated games, players select their strategies according to probabilities. In the well-known rock-paper-scissors game, we use a 3 × 3 payoff matrix. If each player applies the rock, the paper, and the scissors strategy exactly with the probability of 1/3, then the players are in Nash equilibrium. We call this a mixed strategy. That means, if a Nash equilibrium in perfect strategies cannot be found, the solution can be found in a mixed strategy.

The term *Pareto optimal* is also important for social dilemmas. People are in a *Pareto optimal* position, if these positions are on average best for the whole group. However, if one person deviates from his/her position in order to get a better individual payoff, at least one other person of the group will suffer and receive a worse payoff. In other words, no player can improve his/her payoff at the cost of at least another player.

By having defined the basics, we can discuss some practical social dilemmas.

Chapter 6
Social Dilemmas and Problems of Social Order

We describe problems of social orders, and we discuss whether these problems will emerge during the cooperation of digital selves.

It is a fatal mistake to think that irrational results of social problem solutions lead back to irrational behaviors. In a prisoner's dilemma situation, rational-acting persons produce irrational results. In a social dilemma situation, the Nash equilibrium is not Pareto optimal, or a Pareto optimal situation is only achievable if coordination is allowed. We present this situation in the following examples.

In the chicken game, two persons pilot their cars on collision course. The one who avoids the collision first is the looser. If nobody avoids the collision, then both are dead or highly injured. If both persons try to avoid the collision, then nobody has lost his face. The worst situation happens if both stay on course. The corresponding normal form of the game is presented in Fig. 6.1.

Two Nash equilibriums can be identified, namely, the one with the payoffs (2,4) and (4,2). The equilibrium "avoid collision" and "not avoid collision" favors the column player, and the equilibrium "not avoid collision" and "avoid collision" favors the line player. But if both players select the favorable strategy "not avoid collision," then a catastrophe is inevitable.

The famous chicken game can be mapped onto international conflicts as presented in Fig. 6.2 for the Iran conflict. The situation can only be solved through negotiations in order to find a compromise, which is not a Nash equilibrium.

The next example is an N person prisoner's dilemma. Let us have a look at the following decision situation. N persons can decide between the strategies S1 and S2. The following payoffs will be given to people selecting S1 respectively S2:

$$S1 = 2x$$

$$S2 = 3x + 3$$

S1 and S2 are the strategies available for a player; x is the number of players who are selecting strategy S1. In summary, we have N players.

© The Author(s) 2022
F. Barachini, C. Stary, *From Digital Twins to Digital Selves and Beyond*,
https://doi.org/10.1007/978-3-030-96412-2_6

	Avoid collision	Not avoid collision
Avoid collision	3,3	2,4
Not avoid collision	4,2	1,1

Fig. 6.1 Chicken game payoff matrix

	Stop nuclear research program	Continue nuclear research program
Negotiate, no attack on Iran	Compromise	Victory Iran, Defeat Allies
Attack on Iran	Victory Allies, Defeat Iran	War?

Fig. 6.2 International conflict matrix

	S1	S2
S1	4,4	2,6
S2	6,2	3,3

Fig. 6.3 Two players of the N-players game

Let us assume that the group consists of 100 players and 40 of them selected strategy S1. Then they get 80 points and the 60 other players get 123 points. But if all players would have selected S1, then they all would have gotten 200 points. In this case, S1 is the cooperative strategy, and S2 is the defecting strategy. Where is the break-even point? The break-even point is reached when less than 66 people are selecting S1. That means that at least 45 people select S2. In this case, all get less than 200 points.

Thus, a few people can destroy the fruits of the work of the whole group, and if these few people are more than 44 out of 100, then they even get less than they would have achieved through total cooperation. In fact, the best result is achieved if 99 persons select strategy S1. In this case, the defector achieves 300 points and the rest of the group only 198. Which of the two strategies is Nash and which one Pareto? Let us look at the payoff matrix if only two people are playing the game (see Fig. 6.3).

We see clearly that S2 is the Nash equilibrium which yields the payoff (3,3) for both players. If they both would have selected the S1 strategy, which is a Pareto optimum, they would be better off. However, by selecting the S1 strategy, there is the danger that one of them (the defector) selects S2. In this case, the defector gets the maximum payoff (6), and the co-operator gets the minimum (2). This is again a typical prisoner's dilemma.

The above-described game can be generalized to public good games which are typical for social dilemmas. The difference between a public good game and an

N-players game is that in a public good game, a player can decide how much he/she is willing to cooperate.

The problem of organ donations is a typical public good game. Everybody wishes that the collective good organ is available but not many people possess an organ donation pass. If all people follow the S1 strategy which means cooperation, the benefit is maximal for all. Nonetheless, there is a minority contributing and holding an organ donation pass. The solution for such problems is to find incentives so that people switch from the S2 strategy to the S1 strategy. One solution would be that owners of an organ donation pass are preferred when they need an organ transplantation. Other examples are vaccine passes as used during the COVID crisis where people owing such a pass witness less hurdles in daily activities such as traveling or shopping.

There are many more examples of public good games, such as voter turnout rates in democracies, joining of non-profit associations, joining protest demonstrations, contributing to any kind of common group results, or keeping a clean environment where incentives for e-cars in the form of tax reductions are applied or for big industries carbon trading certificates are provided. Incentives try to dissuade people and companies from defection so that Pareto optima can be achieved. By these means, social dilemmas are mitigated.

Since depletion of resources will always result in some kind of social dilemmas, we have to ask ourselves if digital selves might be confronted with the same sort of social order problems. There are two answers to that question.

The first answer is that digital selves should act according to the game theory. This would be an emotion-free, cold world where only utility functions and decisions based on statistics are ruling the behavior of selves.

The second answer is that digital selves and humans will coexist for a long time. Therefore, for the public benefit, digital selves will have to adopt the human way of thinking and behaving. For instance, in a mixed traffic, it makes no sense that autonomous cars follow their own learned algorithms ignoring human driving habits. Thus, digital selves need to balance the collective perspective of humans and trans-humans. Until recently, we designed human-centric AI systems. For the future, we will probably need humanity-inspired AI systems incorporating cultural norms, values, beliefs, and all emotional aspects of human life, so that humans and digital selves can coexist. From this, expectation follows that humans and digital selves represent each other's context which may require representation and intervention in the course of adaptation and collaborative action.

Chapter 7
Emotional Modeling with Spatial Games

Elster (1996) argues that social norms play a role in the generation of emotions. Indeed, emotions are regulated by social norms. Appraisal theory (Frijda, 1986; Lazarus, 1991) explains the role of social norms in the generation and regulation of emotions.

The theory argues that emotion arises from the appraisal process and from the subsequent coping process. During the appraisal process, a person assesses his/her relationship with the environment. Several so-called appraisal variables are individually calculated. The variables serve as an intermediate description of the person-environment relationship. Coping determines how people respond to the appraised significance of events. Problem-focused coping strategies attempt to change the environment, whereas emotion-focused coping strategies alter the mental stance.

The effect of both strategies is a change in a person's interpretation of his/her relationship with the environment. Staller and Petta (2001) integrated an appraisal process into their Tabasco architecture. By using SOAR as a vehicle, Gratch et al. (2006) provided a model of the processes underlying cognitive appraisal. Both approaches are based on a rule-based paradigm. The developers admit that the largest deficiency of their model concerns the impoverished social reasoning.

Jealousy has not been modeled explicitly by other researchers. They have focused on other emotions so far (Picard, 1997; Castelfranchi, 2000).

We model jealousy as one single variable by using a game theoretical approach. We assume that this single variable has been generated either from an antecedent appraisal process or it is given a value which can continuously change in a simulation process. Our agents only comprise one learning function. They are not modeling mind on an individualistic micro-level as demanded by Castelfranchi (2006). However, in Chap. 4, we do present a generic layered incremental architecture, so that the impact of several parallel emerging emotions to collective behavior can be simulated. Game theory, as explained in one of the previous chapters, is well suited to model the intra-relationships between agents.

Emotions play a big role during cooperation. Jealousy seems to play a special role (Barachini, 2015; Brockner, 1988; Pelham & Swann, 1989; Rydell & Bringle,

© The Author(s) 2022
F. Barachini, C. Stary, *From Digital Twins to Digital Selves and Beyond*,
https://doi.org/10.1007/978-3-030-96412-2_7

2007). Just during the COVID-19 pandemic crisis where common goods, such as vaccines, are rare for some parties, we witness a struggle between different social groups and even nations. Jealousy is the main driver in such socioeconomic conflicts. We investigate the impact of jealousy on cooperation, and we simulate the deterministic evolutionary dynamics of populations which are arranged in a lattice.

We concentrate our effort on the question how jealous agents can invade co-operators and vice versa. In this context, we are interested about the impact of rewards to the behavior of agents. We show how complex emergent patterns are generated in certain parameter regions, when spatial cooperation in finite populations is applied.

7.1   The Notion of Jealousy

Cooperation through information sharing is substantially influenced by an individual's level of jealousy. This concept so far has been discussed most often in the context of close relationships. This emotion also plays an important role in the work context. The loss or merely the perceived threat of a loss involves the perception of a rival's intrusion. This rival has the potential to reduce one's self-esteem or undermine a valued relationship. It is hence a characteristic of jealousy that it is triadic, involving the focal individual, the rival, and the valued target person.

Rydell and Bringle (2007) distinguish two types of jealousy. *Reactive jealousy* refers to the emotional components of jealousy. *Suspicious jealousy* contains cognitive and behavioral components. They argue that manifestations and antecedents differ for these two types. Specifically, reactive jealousy that occurs after a major jealousy-evoking event and is thus more closely related to exogenous factors appears to be positively related to trust and negatively related to chronic jealousy. However, suspicious jealousy that occurs before major jealousy-evoking events have occurred is more closely related to endogenous factors and is positively associated with insecurity and chronic jealousy and negatively associated with self-esteem, i.e., an individual's self-evaluation of its own competencies (Rosenberg, 1965), including an affective component of liking itself (Pelham & Swann, 1989). It has been found to positively influence employee attitudes, such as job satisfaction, motivation, and performance (Brockner, 1988).

We assume that suspicious jealousy plays an important role in collaboration by mediating the relationship between self-esteem and the willingness to share knowledge with a person who is a potential rival such that the less self-esteem an individual has, the more they may show suspicious jealousy toward potential rivals, and the less they will be willing to share their knowledge with those. Reactive jealousy, in turn, may negatively influence cooperative behavior only after the occurrence of an event that has evoked an individual's jealousy.

In our simulations, we investigate suspicious jealousy for two reasons. Firstly, it is easy to implement because we use simple matrices for payoff calculations, and secondly, reactive jealousy would impose much more complexity, since we would have to construct specific artificial social injustice examples. However, in real life, reactive jealousy is much better identifiable than suspicious jealousy. Reactive jealousy can easily be detected due to its exogenous nature. The detection of suspicious jealousy needs deeper psychological investigations.

7.2 Cooperation Methods and Modeling: State of the Art

Cooperation implies that a donor pays a cost and a recipient gets a benefit in the short term. In the long term, the relation should be balanced. However, cooperation is always vulnerable to exploitation by defectors. Researchers identified well-established cooperation rules, such as kin selection and related schemes (Hamilton, 1964a, 1964b, 1996, 2001), graph and group selection and related schemes (Ohtsuki & Iwasa, 2004), indirect reciprocity, and spatial evolutionary game theory. Among others, punishment and cooperation were investigated by Sigmund et al. (2001) with prisoner's dilemma games.

In practice, most of the updating procedures in prisoner's dilemma games (Rapoport, 1999) use short memory strategies that utilize only the recent history of previous interactions. Li and Kendall (2014) prove that longer memory strategies outperform shorter memory strategies statistically in the sense of evolutionary stability. In another recent publication, Lu et al. (2018) show that long memory effects can change the cooperation behavior in the spatial prisoner's dilemma game. Chen et al. (2016) studied the impact of reputation on evolutionary cooperation. Both approaches apply Fermi rules for the updating procedures and belong to the class of long-term memory strategies.

However, Conte et al. (1999) proved that prisoner's dilemma games (Axelrod, 1984) cannot account for a theory of cooperation, since bargaining is not possible. Therefore, in a previous research, we applied an inter-disciplinary approach based on the business transaction theory of Barachini (2007), which will be explained in Chap. 5 in detail. We underpin the theory in a broad industry survey (150 European companies, 3 countries) (Barachini, 2009). Similar to Akerlof (1970), Spence (1973), and Stiglitz (1975), the theory is based on the idea that information is asymmetric in its nature.

Although Conte et al. criticized the approach of using prisoner's dilemma games because bargaining is not possible, we apply a game theoretical approach based on a short-term memory strategy. We compare this deterministic spatial approach with the stochastic one in Chap. 5.

7.3 Methodology

In the stochastic approach, members of a population meet each other randomly. In the spatial approach, members of a population are arranged in a two- or higher-dimensional array. In each round, every individual agent, represented by a cell, plays a game with its immediate neighbors. After the game, each cell is occupied by its original owner or by one of the neighbors, depending on who scored the highest payoff in that round.

On a spatial multi-dimensional grid, each agent occupies a position on the grid, the cell, and interacts with all of its neighbors. In normal life, an agent can select from an infinite number of strategies. In a deterministic spatial game, each agent adopts the strategy with the highest payoff in its neighborhood. The agents are updated in synchrony.

The fate of a cell depends on its own strategy, the strategy of its neighbors, and the strategy of their neighbors. If we consider a three-dimensional cube containing 5×5 mini cubes per surface, and we observe the inner cube, then the fate of this inner cube depends on its own strategy, the strategies of its 26 neighbors ($3*3*3-1$), and the strategies of their 98 neighbors (the outer layer of the $5*5*5$ cube). Thus, 125 cells in total determine what will happen to the inner cell of a cube.

To make it easier for the reader, we show experiments with two-dimensional arrays only. In this case, each cell has eight direct neighbors because we apply the Moore neighborhood. Thus, 25 cells in total determine what will happen to a cell. To repeat, an agent represented by a cell will retain its current strategy, if it has a higher payoff than all of its neighbors. Otherwise, the agent will adopt the strategy of that neighbor with the highest payoff. We have to admit that in the simulated environment, we have perfect knowledge about the characteristics of an agent. In real life, that sort of information is difficult to discover. For the software implementation, we used an external module known as "VirtualLabs," which is publicly available at the University of Vienna (https://www.univie.ac.at/virtuallabs/).

In our examples, we simulate agents with only three characteristics. That means an agent can choose between three strategies. We have defectors (D), co-operators (C), and conditional co-operators (CC). The latter represent jealous co-operators. The level of jealousy is determined by α respectively by γ of the payoff matrix in Fig. 6.1. Both values vary between 0 and 1. The higher both values are, the higher the probability of cooperation, and the lower the jealousy factor. Consequently, an agent can choose between the three strategies C, D, and CC. The numbers in the following payoff matrices preceding the colon represent the payoff given to the line player, and the numbers after the colon represent the payoff given to the column player.

The payoff matrix of Fig. 7.1 can be interpreted as follows: If two co-operators interact, then both receive one point. If a defector meets a co-operator, the defector gets a payoff a > 1, and the co-operator gets the payoff 0. The interaction between two defectors leads to a very small positive payoff ε, for both. If a co-operator meets a conditional co-operator, then the co-operator gets payoffs between 0 and 1, and the

	C	D	CC
C	1, 1	0, a	$\gamma, \gamma + (1-\gamma)a$
D	a, 0	ε, ε	$\gamma a + (1-\gamma)\varepsilon, (1-\gamma)\varepsilon$
CC	$\alpha + (1-\alpha)a, \alpha$	$(1-\alpha)\varepsilon, \alpha a + (1-\alpha)\varepsilon$	$\gamma\alpha + (1-\alpha)\gamma a + (1-\alpha)(1-\gamma)\varepsilon,$ $\gamma\alpha + (1-\gamma)\alpha a + (1-\alpha)(1-\gamma)\varepsilon$

Fig. 7.1 Payoff matrix

	C	D	CC
C	1, 1	0, a	$\gamma, \gamma + (1-\gamma)a$
D	a, 0	0, 0	$\gamma a, 0$
CC	$\alpha + (1-\alpha)a, \alpha$	$0, \alpha a$	$\gamma\alpha + (1-\alpha)\gamma a,$ $\gamma\alpha + (1-\gamma)\alpha a$

Fig. 7.2 Simplified payoff matrix

conditional co-operator gets payoffs between 1 and a, with a >1. If a conditional co-operator meets a defector, then the conditional co-operator gets payoffs between 0 and ε, and the defector gets payoffs between ε and a. When two conditional co-operators meet, then they both have a payoff range in between 0 and a, a > 1.

The distribution of the rewards of the agents correlates to the prisoner's dilemma in which two agents try to maximize their payoffs by cooperating with or betraying the other agent (Rapoport, 1999). As discovered by the biologists Maynard-Smith and Price (1973), such distributions are also a common phenomenon in the logic of animal conflicts.

In case of setting α to 1 or 0 and setting γ to 1 or 0, we get the simple matrix consisting only of co-operators (C) and defectors (D). Since we simulate jealousy by the parameters α and γ, more jealousy means less cooperation, we have to keep the matrix above, but for simplicity we choose to set $\varepsilon \rightarrow 0$. In fact, this simplification of the matrix in Fig. 7.1 yields the matrix of Fig. 7.2, which represents a prisoner's dilemma. To explore different evolutions of populations, we vary the parameters a, α, and γ. We perform our experiments by investigating suspicious jealousy only, because we would have to construct specific artificial social injustice examples, so that reactive jealousy could be investigated. This would make our simulations even more complex.

As explained, modeling of suspicious jealousy in the deterministic spatial approach is performed by varying the parameters α and γ. Since their values vary in between 0 and 1 with a > 1, this is a large mathematical space, and our simulation results fill more than thousand books. Therefore, we present an extract of all possible outcomes by discussing some interesting parameter regions, especially those which are represented by an iterated prisoner's dilemma (IPD). In the IPD game, two agents choose their mutual strategy repeatedly. They also have memory of their previous behaviors and the behaviors of their opponents. The payoff matrix of Fig. 7.2 yields an IPD only, if certain conditions are fulfilled.

As explained by Rapoport (1999) and Chong et al. (2007), the reward of the co-operators, in our example the value 1, needs to be higher than half of the sum of the sucker's payoff and of the temptation value. In other words, "$1 > (a + 0)/2$" and "$1 > (\alpha + (1-\alpha) a + \alpha)/2$." Thus, $a < 2$ and $\alpha < 1$, respectively $\gamma < 1$. These additional conditions exclude that the alternate strategy combinations "CD, DC, CD, DC..." yield better results than continuous cooperation. These conditions are set to prevent any incentive to alternate between cooperation and defection. In the subsequent simulations, we take these conditions into consideration. Therefore, our examples are of type IPD.

Our update rules are based on deterministic dynamics. Each cell in the lattice is given to whoever has the highest payoff in the direct neighborhood, and all cells are updated in synchrony. This approach allows us to study the properties of deterministic spatial dynamics in discrete time. Although the fate of each cell is deterministic, the overall population dynamics can be richer than in stochastic approaches.

We did not investigate asynchronous updating procedures, where if one cell is chosen at random, its own payoff and the payoffs of all neighbors are determined. Then the individual cell is updated. Asynchronous updating means overlapping generations; synchronous updating means no overlapping generations. Asynchronous updating introduces random choice. The synchronous approach will exactly reproduce the population dynamics when repeated with the same initial conditions.

We are investigating the influence of jealousy on cooperation in a spatial IPD by using a short memory updating procedure without statistical means. Subsequently, we show under which conditions co-operators can invade defectors and vice versa. Individual agents of the population are placed in a two-dimensional lattice. Throughout the rest of part II of this book, we apply the matrix of Fig. 7.2 in order to calculate the fate of the cells.

7.4 Co-operators Invaded by Defectors

We investigate the conditions for a single defector to invade a population of co-operators. Note that a co-operator could be of type C (unconditional cooperation) or CC (conditional cooperation, based on the level of suspicious jealousy).

Figure 7.3 shows the conditions for a single defector to invade a population of jealous co-operators. We set the level of suspicious jealousy to 10%, which means the cooperation levels of α and γ are set to 0.9. Based on the payoff matrix from Fig. 7.2, we can evaluate the payoff of the single defector (red cell). The single defector is surrounded by eight jealous co-operators (blue cells). Therefore, the payoff of the defector is eight times the D-CC combination ($8 \times \gamma \times a$) plus one D-D combination. Since the D-D combination yields 0, we get $8 \times 0.9 \times a = 7.2\ a$. Therefore, the red cell has payoff 7.2 a.

6,48 + 0,72 a	6,48 + 0,72 a	6,48 + 0,72 a	6,48 + 0,72 a	6,48 + 0,72 a		
6,48 + 0,72 a	5,67 + 0,63 a	5,67 + 0,63 a	5,67 + 0,63 a	6,48 + 0,72 a		
6,48 + 0,72 a	5,67 + 0,63 a	7,2 a	5,67 + 0,63 a	6,48 + 0,72 a		
6,48 + 0,72 a	5,67 + 0,63 a	5,67 + 0,63 a	5,67 + 0,63 a	6,48 + 0,72 a		
6,48 + 0,72 a	6,48 + 0,72 a	6,48 + 0,72 a	6,48 + 0,72 a	6,48 + 0,72 a		

Fig. 7.3 Payoffs of a single defector cell and of the surrounded jealous co-operator cells

6,48 + 0,72 a	6,48 + 0,72 a	6,48 + 0,72 a	6,48 + 0,72 a	6,48 + 0,72 a	6,48 + 0,72 a	6,48 + 0,72 a
6,48 + 0,72 a	5,67 + 0,63 a	4,86 + 0,54a	4,05 + 0,45 a	4,86 + 0,54 a	5,67 + 0,63 a	6,48 + 0,72 a
6,48 + 0,72 a	4,86 + 0,54 a	4,5 a	2,7 a	4,5 a	4,86 + 0,54 a	6,48 + 0,72 a
6,48 + 0,72 a	4,05 + 0,45 a	2,7 a	0	2,7 a	4,05 + 0,45 a	6,48 + 0,72 a
6,48 + 0,72 a	4,86 + 0,84 a	4,5 a	2,7 a	4,5 a	4,86 + 0,84 a	6,48 + 0,72 a
6,48 + 0,72 a	5,67 + 0,63 a	4,86 + 0,54 a	4,05 + 0,45 a	4,86 + 0,54 a	5,67 + 0,63 a	6,48 + 0,72 a
6,48 + 0,72 a	6,48 + 0,72 a	6,48 + 0,72 a	6,48 + 0,72 a	6,48 + 0,72 a	6,48 + 0,72 a	6,48 + 0,72 a

Fig. 7.4 9D square

The value of the blue cell above the red cell is calculated as follows: the blue cell above the red cell is a jealous co-operator, and it is surrounded by seven other blue cells and one red cell. Therefore, the payoff of the jealous co-operator is seven times the CC-CC combination [7 × (0.9 × 0.9 + 0.1 × 0.9 × a)] plus one CC-D combination. Since the CC-D combination yields 0, we get "5.67 + 0.63 a" for this blue cell. Similarly, all the other cells of Fig. 7.3 are filled in by using the payoff matrix of Fig. 7.2.

If "7.2 a > 6.48 + 0.72 a," which means a > 1, then the defector will take over all its neighbor cells. This will then yield Fig. 7.4.

In Fig. 7.4, the central defector has a payoff of zero, and the rest is evaluated using our payoff matrix from Fig. 7.2.

- The 9D square defector cells will turn into a single defector again, if "4.86 + 0.54 a > 4.5 a." Thus, if "1.22 > a," then Fig. 7.4 will again turn into Fig. 7.3.
- The 9D square stays stable, if "6.48 + 0.32 a > 4.5 a > 5.67 + 0.63 a." Thus, if "1.71 > a > 1.46," then the configuration of Fig. 7.4 will not change.
- There will be further growth of the 9D square, if "4.5 a > 6.48 + 0.32 a." Thus, if "a > 1.55," we have further linear growth of the 9D square to a 25D square, and so on, until eternity.
- Fig. 7.4 will turn into the cross of Fig. 7.5, if "5.67 + 0.63 a > 4.5 a > 4.86 + 0.54 a." Thus, if "1.46 > a > 1.22," we get a cross, which will turn again into a single defector of Fig. 7.3. In this case, we have a period two oscillator.

The above example is only one of many million, and it shows one deterministic pattern. Many more deterministic patterns have been observed. We also simulated jealous co-operators invading defectors. It is strongly dependent on the value of a (the reward) and the values of γ and α, whether the world is dominated by defectors

	6,48 + 0,72 a	5,67 + 0,63 a	5,67 + 0,63 a	5,67 + 0,63 a	6,48 + 0,72 a	
	5,67 + 0,43 a	4,05 + 0,45 a	4,5 a	4,5 + 0,45 a	5,67 + 0,63 a	
	5,67 + 0,63 a	4,5 a	3,6 a	4,5 a	5,67 + 0,63 a	
	5,67 + 0,63 a	4,05 + 0,45 a	4,5 a	4,05 + 0,45 a	5,67 + 0,63 a	
	6,48 + 0,72 a	5,67 + 0,63 a	4,05 + 0,45 a	5,67 + 0,63 a	6,48 + 0,72 a	

Fig. 7.5 Cross

				6,48 + 0,72 a		
		4,05 + 8,45 a	4,86 + 0,54 a	5,67 + 0,63 a		
		2,7 a	4,5 a			
		0				

Fig. 7.6 Conditions for fractals and kaleidoscopes

or by co-operators or whether there is a dynamic balance. In certain parameter regions, the abundance of jealous co-operators depends on the starting configuration of the grid. In other parameter regions, the frequency of co-operators is almost constant, especially in very large arrays.

The simulations are all based on finite automata. There is one very interesting constellation, where dynamic kaleidoscopes and fractals are generated. The initial conditions for these kaleidoscopes are shown in Fig. 7.6.

- If "4.5 a > 6.48 + 0.72 a," then defectors win at corners. That means a > 1.71.
- If "2.7 a < 4.05 + 0.45 a," then jealous co-operators win along lines. That means a < 1.8.
- Thus, if "1.8 > a > 1.71," then we have a clash of extremes.

Figures 7.7 and 7.8 show the fractals after 124 and after 128 rounds for the parameter a, using a jealousy factor of 10% for each co-operator on a 128 × 128 spatial grid. We started with a single defector.

Only by coloring the grid, the pattern can be optically observed:

- Yellow is a defector that was a jealous co-operator in the previous round.
- Green is a jealous co-operator that was a defector in the previous round.
- Red is a defector that was a defector in the previous round.
- Blue is a jealous co-operator that was a jealous co-operator in the previous round.

In Figs. 7.7 and 7.8, the fractals repeat themselves at the power of 2. It can be observed that the number of defectors converges toward 2/3. This can be observed after many thousand rounds. The jealous co-operators cannot be totally invaded by defectors in this case.

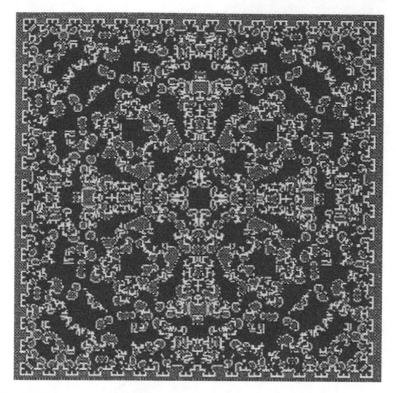

Fig. 7.7 Kaleidoscope after 124 update rounds

7.5   Defectors Invaded by Co-operators

For "1.8 > a > 1.71," we show in Fig. 7.9 a fractal after 260 rounds, using a jealousy factor of 50% for each co-operator. We used again a 128 × 128 spatial grid, representing a population size of 16,384 cells. We started with a 9D block of co-operators within defectors:

- White is a defector that was a jealous co-operator in the previous round.
- Blue is a jealous co-operator that was a defector in the previous round.
- Red is a defector that was a defector in the previous round.
- Dark is a jealous co-operator that was a jealous co-operator in the previous round.

In Fig. 7.9, the fractals repeat themselves at the power of 2. It can be observed that the number of defectors converges toward 2/3. The defectors cannot be invaded by jealous co-operators. With the given jealousy factor of 50% for each co-operator and a reward range between 1.71 and 1.80, defectors cannot be invaded. Thus, the strategy of the defectors is evolutionary stable. In terms of Maynard-Smith, both special cases of Figs. 7.8 and 7.9 represent a cyclic evolutionary stable strategy (ESS).

Fig. 7.8 Kaleidoscope after 128 update rounds

7.6 Findings

In this chapter, we investigated update procedures based on deterministic dynamics, in a spatial grid. If update rules are determined by deterministic dynamics, then we can even produce fractals and kaleidoscopes, which repeat themselves at the power of 2 and where all types of cells can find their place. Thus, jealous cells are not eliminated under specific parameter conditions. This is a special case of equilibrium.

We conclude that under special constellations of the cells, the presented spatial game theoretical approach has a high variety in dynamical behaviors. This variety mainly depends on the payoff function, which is determined by the intensities of the reward and of suspicious jealousy, and a very specific constellation of edge and line cells in the grid. Whether less or more reward is needed depends on the jealousy level. The parameter regions are shifted accordingly.

If the level of jealousy would be determinable in real life, then we could make estimations of the collective behavior of groups, and even then, it could be chaotic. But is a chaotic behavior dangerous for groups and their managers? Yes and no! Kaleidoscopes indeed make the impression of chaos, but in the long run, the numbers of co-operators and defectors stay balanced. In spite that jealous

Fig. 7.9 Fractal after 256 update rounds

co-operators are switching continuously from one cell to the next, the equilibrium of the group might not be endangered. Therefore, the other parameter regions could be more relevant for real-life investigations, since they can tip populations completely, as it was explained in Fig. 7.4 for the 9D square, when the reward is higher than 1.55.

We simulated suspicious jealousy by simply modifying the two parameters α and γ. The level of jealousy and the amount of the reward influence social behavior and group structures, as it has been shown in the colored pictures.

We suspect that behavior of humans in groups is still more complex, since beside egoism, trust, and jealousy, additional emotions, such as joy, fear, surprise, anger, or anticipation, need to be considered. It is very hard to control emotions during experiments with humans and to identify root-cause effects triggered by these emotions. Moreover, in reality, it is probably impossible to identify people and measure their exact levels of suspicious jealousy. This type of jealousy can be regarded as a typical disease pattern. In order to establish unambiguous psychological experiments with humans, all these facts need to be considered.

In real life, a mixture of stochastic and spatial interactions occurs. In highly automated companies, with fixed defined processes and few customer contacts, spatial internal interaction in the production line is probably dominant. In random crowds, stochastic interaction is dominant. If, for spatial interaction, we decide to equip artificial agents with emotions such as jealousy, then the dynamic behavior of the crowd will correlate with the emergent presented patterns.

In the next chapter, we model agent-based stochastic interactions. We present a generic-layered architecture which is capable to model emotions such as jealousy, joy, fear, surprise, anger, anticipation, etc. in parallel.

Chapter 8
Agent-Based Stochastic Simulation of Emotions

In this chapter, we model jealousy of agent-based stochastic interactions from a very different perspective. We study the impact of jealousy to cooperation models by using dynamic artificial agents and agent populations following combined characteristics. We simulate a dynamic multi-player public goods game, and we use the Business Transaction Theory (Barachini, 2007) as a basic layer for information valuation.

Our approach enables us to simulate small to large populations in a short time, and it enables flexibility in changing agent characteristics. The technical framework is based on the Eclipse development environment. The artificial agents are programmed in Java.

Since we simulate human properties, we need to understand the basic characteristics of cooperation. In knowledge management, exchange and sharing are the activities that characterize knowledge-driven collectives. More precisely, when we talk about exchange and sharing, we talk about information transfer between humans.

Since repeated prisoner's dilemma games fall short for modeling truly cooperative behavior (Conte et al., 1999) because the possibility of bargaining is inexistent, we use, in addition to evolutionary game theory, artificial agents for social simulation. The approach combines the advantages of several disciplines, namely, artificial intelligence, psychology, and economy.

8.1 The Economic Perspective of Communication

In cooperation, a donor pays a cost and a recipient gets a benefit. However, cooperation is always vulnerable to exploitation by defectors. Researchers identified well-established cooperation rules, such as kin selection and related schemas (Hamilton 1964a, 1964b, 1996, 2001), graph and group selection and related schemas (Ohtsuki & Iwasa, 2004), indirect reciprocity, and spatial evolutionary game theory.

© The Author(s) 2022
F. Barachini, C. Stary, *From Digital Twins to Digital Selves and Beyond*,
https://doi.org/10.1007/978-3-030-96412-2_8

Among others, punishment and cooperation were investigated by Sigmund et al. (2001). As explained above, Conte et al. (1999) proved that prisoner's dilemma games cannot account for a theory of cooperation since bargaining is not possible.

The research results gathered from the different disciplines encourage the application of an inter-disciplinary approach. To investigate cooperation methods from an information economics perspective, we developed the *Business Transaction Theory*, which relates information exchange to modern portfolio theory. We underpin the theory in a broad industry survey (150 European companies, 3 countries) published in JKM (Barachini, 2009). Similar to Akerlof (1970), Spence (1973), and Stiglitz (1975), the theory is based on the idea that information is asymmetric in its nature.

Akerlof (1970) experimented with car markets and the asymmetric information situation between the seller, who knows everything about the car, and the buyer, who knows only what the seller has told him. The fact that the potential buyer is aware of his lack of information leads to the default assumption of the buyer that the used car is a "lemon" (a car of low quality). Therefore, the potential buyer will always bargain for a lower price, which drives high-quality cars out of the market so that finally only "lemons" will be offered. This process is called "adverse selection." Akerlof observed that this effect can be weakened by factors such as "repeated sales" and "reputation," which, however, is less true for the insurance, labor, and credit markets. Spence's job market studies (Spence, 1973) yielded similar results.

Stiglitz took the work of Spence as a basis and focused especially on the insurance market. He introduced the concept of "screening," which is a self-selection mechanism, where agents are offered a variety of contracts. The selections done by the agents lead to a revelation of their risk level, as risk averse agents tend to choose contracts which charge lower premiums, but higher deductibles. According to Stiglitz, the Pareto-optimal outcome would be a full screening that identifies each agent's true capabilities, but it cannot be seen as market equilibrium due to missing sustainability.

The theories of Akerlof, Spence, and Stiglitz focus on specific markets. Market participants exploit asymmetric information to gain profits. The market of the Business Transaction Theory is characterized by information itself. No specific market is necessary. Thus, it is the basic underlying information exchange mechanism between humans that is characterized by the Business Transaction Theory, independent from any specified markets. We apply this theory as the basic communication mechanism also for agents.

8.2   Business Transaction Theory

The Business Transaction Theory defines two types of information exchange: Type-1 is the immediate exchange of information in both directions. Thus, donor and recipient both provide information. This type of duplex information provision can be mapped to over-the-counter (OTC) businesses of banks. The difference to the well-known OTC business is that there are no intermediaries involved such as dealers and

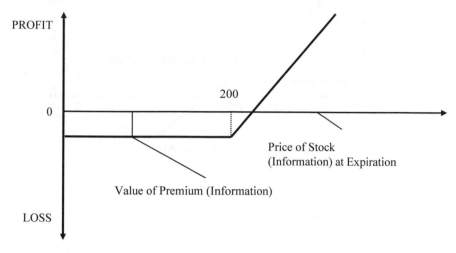

Fig. 8.1 P&L graph for "buy a call"

brokers because the so-called information market is generated implicitly in the minds of the parties involved.

Type-2 of information exchange is more complicated because information flow is unidirectional at first. This is the case when we offer information to individuals in the hope to get even more valuable information back some day in the future. Type-2 of information exchange can be mapped to options.

In the investment world (Sharpe et al., 1995), an option is a type of contract between two people where one person, the writer, grants the other person, the buyer, the right to buy a specific asset at a specific price within a specific time. Alternatively, the contract may grant the other person the right to sell a specific asset. The variety of contracts containing an option feature is enormous.

Type-2 of information exchange can be mapped to the call option for stocks. It gives the buyer the right to buy a specific number of shares of a specific company from the option writer at a specific purchase price at any time[1] up to a specific date. Figures 8.1 and 8.2 show the profit and loss (P&L) graphs of a buyer and a seller. The buyer of a call option will have to pay the writer a premium in order to get the writer to sign the contract. The *fair value* of an option can be evaluated using the binomial option pricing model or the more recent method by Black-Scholes (in Sharpe et al., 1995):

$$\text{Fair value} = N(d1) \times Ps - E \times N(d2)/e^{RT}$$

---

[1] For US options only.

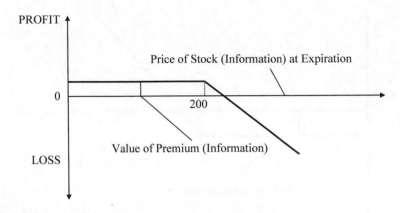

Fig. 8.2 P&L graph for "write a call"

where:

d1: $(\ln(Ps/E) + (R + 0.5\ \sigma^2)T)/\sigma \times sqrt(T)$
d2: $d1 - \sigma \times sqrt(T)$
Ps: Current market price of underlying stock
E: Exercise price of option
N: Normal
R: Compound risk-free rate of return
σ: Risk of the underlying stock
sqrt: Square root
T: Time remaining until expiration

　　Figure 8.1 relates the value of a call option with an exercise price of 200 to the price of the underlying stock of expiration. If the stock price is below 200, the option will be worthless when it expires. In this case, the writer will gain (see Fig. 8.2) the premium. If the price is above 200, the option can be exercised for 200 to obtain a security with a greater value, resulting in a net gain to the option buyer that will equal the difference between the securities market price and the 200-exercise price. However, margin requirements, commission payments, and other market-making activities make the procedure more complicated in practice.

　　In the Business Transaction Theory, Type-2 of information exchange means that one person (the buyer) provides information and hopes to get even more valuable information in the future (see Figs. 8.1 and 8.2). The information offered to the writer has some value—the premium. The buyer hopes to get back another type of information that is at least as valuable as the premium. Thus, the underlying traded goods are not stocks, but information.

　　In the case of the classical stock market, the Black-Scholes formula is based on statistics, the exercise price is known, the risk of the underlying common stock can be evaluated, and the option has a well-defined expiration date.[2]

---

[2] This is true for European options—US options can be exercised arbitrarily.

In the case of information brokerage, it is more difficult to evaluate a fair price for a piece of information since we do not even know the value of the underlying good because it is an unknown piece of information which might, or might not, be offered by the writer sometime in the future. In the Black-Scholes formula, the current market price of the underlying stock can be evaluated. In contrast, no objective evaluation can be performed for information generated by humans since information is always evaluated subjectively. The evaluation function might even be similar, but due to differing context knowledge, the same piece of information may still be evaluated differently on an individual basis. Therefore, statistics as in the Black-Scholes formula cannot be applied immediately in the Business Transaction Theory, since Ps, E, R, and σ now represent subjectively assessed values.

The parameter T is indeterminable since it is not known when or even whether at all valuable information will be received in the future. As a consequence, all parameters have to be simulated in a scientific environment so that the value of the information transferred between agents could be evaluated. Additionally, the social behavior of agents reacting to information needs to be modeled.

8.3 Modeling Approach

We combine cognitive models of social agents with evolutionary game theory. As the basic information exchange mechanism between agents, we use business transactions. In this section, we describe the implementation of the business transaction layer, the research design for simulating jealousy, and the structure of the framework.

8.3.1 *Implementation of the Business Transaction Layer*

We simulate cooperative behaviors of finite populations by simulating the Black-Scholes parameters and by using payoff matrices as they are used in evolutionary game theory. The latter is used to evaluate certain states, i.e., goals of individual agents and crowds; the former is used to simulate the dynamical changing utility of information that is exchanged between agents. For calculating the Black-Scholes parameters, we propose the following procedures:

As explained in the previous section, the utility of information must be calculated or estimated. For communities of practice (CoPs) which are established on the web, Schmidt (2000) proposes two practical approaches in his experiments. He introduces the term "Knowledge Euro" for his approaches. One approach allows the information provider to simply fix a price before the information is delivered. The other leaves it up to the receiver to judge the value of the information. In the latter case, it is up to the receiver to pay a price on a donation basis. In this way, we also consider bargaining, which cannot be modeled in repeated prisoner's dilemma games (Conte et al., 1999).

We propose to calculate the Black-Scholes parameters in the following way:

Ps: The utility of the current market price of underlying information, a future value, can be fixed by the receiver (donation basis).

E: The utility of the exercise price can be fixed by the provider (writer).

T: The time whether and when both agents meet in the future can be simulated through different models, i.e., by simulating a Poisson distribution for meetings. If no meeting takes place at time T, then the call is not exercised (information provision is not reciprocated).

R: The compound risk-free rate can be set to a constant parameter.

σ: The risk of the underlying information can be simulated in different ways.

We can simulate closed marketplaces, where prices behave according to certain functions (logarithmic, linear, scattered, etc.), or we memorize prices fixed by information producers and receivers on donation basis and calculate the variance of these historical data.

Another possibility is the direct calculation of the so-called implicit volatility. This can be achieved in the following way: We put the current market price of the information (either determined by buyer or writer) on the left-hand side of the Black-Scholes formula. Next, all the other factors, except σ, are entered on the right-hand side, and a value for σ, the only unknown variable, is found. This value denotes the implicit volatility of information.

Implicit volatility may be explained in a simple example on a monetary basis: If we assume that the risk-free rate is 6% and that a 6-month call option with an exercise price of 40 € (the writer's estimated utility of future information) sells for 4 € (utility of buyer's information), then the price of the underlying good is 36 € (the writer's real utility of future information). Different estimates of σ can be plugged into the right-hand side of the Black-Scholes formula, until a value of 4 € is achieved. In this example, an estimated value of .4 (40%) for σ will result in a number for the right-hand side of the Black-Scholes formula that is equal to 4 €, the current market price of the call option (the utility of the information provided by the option buyer) that is on the left-hand side. In this example, σ can be estimated for a 6-month option not only for an exercise price of 40 € but also for higher or lower exercise prices. Following Sharpe et al. (1995), after sampling σ values over a sufficiently large range of exercise prices, we can calculate a best estimate for σ as average over these values.

By using payoff matrices similar as described in Chap. 4 and the above-described parameter calculations, the value of an individual's payoff is dependent on the achievement of its goals which in turn can be interpreted as an indicator of its fitness. The same holds for groups.

We simulate the parameters of the Black and Scholes formula as explained above, and we use the Business Transaction Theory (BTT) as the basic evaluation mechanism for rating the information that is exchanged in between artificial agents constituting finite populations of different sizes.

One of the major goals of economic theory has been to explain how cooperation among individuals can happen in a decentralized setting. It was thought that contracts between individuals could suffice to steer their behavior. But the assumption that a contract could completely specify all relevant aspects of a relationship and could be enforced at zero costs to the exchanging parties is not applicable to many forms of cooperation.

Explicit contracts, economic institutions, and also individuals depend on incentive mechanisms involving strategic interaction. However, some research results show that incentives sometimes undermine moral sentiments (Bowles, 2008) and need to be introduced very carefully. Therefore, we concentrate our research toward incentive mechanisms and their use with respect to jealousy. Our agents are able to learn using reinforcement and evolutionary adaptation and therefore can change their behavior so that they can achieve individual or collective goals. We play several rounds with different group size and agent characteristics of a specific multi-player public goods game "establish homes" and evaluate the results with payoff matrices as described in Chap. 4.

An individual agent will exert a significant influence to the other agent's cooperative behavior, in particular on their knowledge-sharing behavior. The more agent characteristics we have, the better we can simulate reality. However, we need to reduce complexity in order to keep our experiments tractable. Therefore, we use jealousy only in a first step. The intensity of jealousy is dynamically changed during the game playing process. The dynamic adaption depends on the success or failure to reach group goals and on incentives such as praise and blame which are optionally injected into the public goods game "establish homes" described subsequently.

8.3.2 Research Design for Jealousy

We model a dynamic multi-player public goods game with a finite population. From one round to the next, we monitor the agent characteristics. The agent characteristics are modified from one round to the next according to specific rules, such as haploid reproduction,[3] or according to reinforced coordination strategies.[4]

Similar to the simulations in Chap. 4, we assume a population of unconditional co-operators, conditional (jealous) co-operators, and egoist agents.[5] Consider a population in which agents can live and work alone or in groups. By the *fitness* of an agent, we mean the expected accumulated knowledge calculated with the

---

[3] As used in biology for inheritance.

[4] A coordination strategy is successful the better the goals could be achieved in previous rounds (higher payoffs).

[5] Alternatively, we can start with unconditional cooperators only and then let dynamically evolve the other two types out of them, dependent on strategies learned in previous rounds during the game.

business transaction function (Black-Scholes) during one round minus the probability that the agent leaves the group.

Agents can also cooperate on a conditional or unconditional basis in a group, each producing an amount b at cost c. All benefits and costs are expressed in fitness units, as calculated by the business transaction function. We assume that the output of a group is shared equally by all members, so if all agents cooperate, each has a net group fitness benefit $b - c > 0$.

Groups consist of three types of actors. The first type, whom we call *jealous co-operators*, works conditionally and cooperates until a certain level of fitness is reached in their counterparts. The second type, whom we call *egoists*, maximizes fitness. They work and cooperate only to the extent that the expected fitness cost exceeds the expected cost of defecting. The third type, whom we call *co-operators,* works unconditionally, and they always cooperate (high agreeability). The agent types can change dynamically from round to round.

Agents change their type with probability $1 - \beta$ from one round to the next. With probability $\beta/2$, an agent takes on each of the other two types. We call β *the rate of change.*[6] Also, with probability $1 - \beta$, egoistic agents inherit the estimate of $s > 0$ (the cost of being eliminated or isolated in the group because of excessive egoism) from the previous round. With probability β, an agent in the next round is a mutant whose s is drawn from a uniform distribution on $[0, 1]$. Thus, s is an endogenous variable.

Jealous co-operators mutate to egoists as soon as they meet agents with sufficiently high accumulated fitness. Suppose an egoist agent defects (i.e., does not work or only gathers information) a fraction Ss of the time, so the average rate of defecting is given by $S = (1 - P*fj - fc)*Ss$, where fj is the fraction of the group, which consists of jealous co-operators, fc is the fraction consisting of co-operators, and P is the probability that a jealous agent cooperates. The fitness value of the group output is $n*(1 - S)*b$, where n is the size of the group.

Since output is shared equally, each member receives $(1 - S)*b$. The loss to the group from an egoist defecting is $b*Ss$. The *fitness cost of the working function,* which can be written as $D(Os)$, where $Os = 1 - Ss$, is increasing and convex in its argument. Expending effort always benefits the group more than it costs the workers, so $Os*b > D(Os)$ for Ss $(0, 1]$. Thus, at every level of effort, Os, working and cooperating helps the group more than it hurts the worker and co-operator.

Further, we assume that the cost of effort function is such that in the absence of jealous co-operators, members face a public goods problem (i.e., an n-player prisoner's dilemma), in which the dominant strategy is to contribute very little or nothing. We model jealousy as follows. The fitness cost of a conditional co-operator to detect egoists is $Cp > 0$. A member defecting at rate Ss will not be selected for cooperation with probability $fj*Ss$. Egoist agents, given their individual

---

[6]The rate of change will be normally determined by the fulfillment of goals in previous rounds calculated with payoff matrices, but in order to keep the example easy understandable, we assume a simple change rate.

assessment s of the cost of being ignored, and with the knowledge that there is a fraction fj of jealous agents in their group, choose a level of defecting, Ss, to maximize expected fitness. Writing down the expected fitness cost of working, FCW (Ss), as the cost of effort plus the expected cost of being ignored, plus the agent's share in the loss of output associated with one's own defection, we get FCW $(Ss) = D(1 - Ss) + s*fj*Ss + Ss*b/n$. Then, egoist agents select Ss, the value which minimizes the fitness cost of working (FCW).

The expected contribution of each group member to the group's population in the next round is equal to the member's fitness minus (for the egoist agent) the likelihood of defection. Different kinds of incentives may be imposed per round on individual agents or on complete groups if more than one group is simulated. However, research results show that incentives may undermine moral sentiments and therefore are producing worse group results as if they would not have been imposed (Falk & Kosfeld, 2006). Both argue that "control aversion" may be a reason that incentives degrade performance. Fehr and Rockenbach (2003) argued that even if incentives reduce the total gains of a group, their use may give the principal a sufficient large slice of the smaller pie to motivate the principal to use them.

By positive incentives, we simply mean the donation of additional scores to the agent in a multi-player public goods game, in reality praise for cooperation. An agent's goal might be to establish a house for himself. He gets a certain amount of scores for reaching the goal. On the other hand, he is getting scores when he helps establish houses for others. The more houses established in the group, the fewer scores the agent might get for his established house, but the more scores will be distributed to the group. Negative incentives[7] will reduce the distribution of scores to individual agents or groups.

As a variation of the experiment "establish homes," we would even be able to impose "signals," as defined by Spence (1973), so that agents can be distinguished by their education level.[8] According to our framework, we hope to simulate various combinations of agent characteristics.

8.3.3 Framework

We can simulate jealousy in isolation, subsequently with signaling and incentive injection. Inside the communication layer in between agents, we use implicit volatility. Later, we can use linear functions. Utility of information is fixed by receiver and provider. This approach shows a clear order of investigation from simple to

---

[7]Punishment, in reality blame for uncooperative behavior.

[8]Different education levels imply different information utilities since higher-skilled agents (people) will deliver more reliable information (if they do not lie or pursue specific strategies).

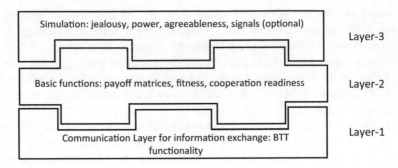

Fig. 8.3 Layer model

medium complexity[9] inside the communication layer as well as in the higher level of the experimental setting. Simulations of more complex agent characteristics might be intractable.

On the highest level (Layer-3), we implement the agent characteristics agreeableness, jealousy, power, and combinations thereof. Optionally, we can impose signaling. On the second level (Layer-2), we implement knowledge payoff (fitness), cooperation readiness (conditional and unconditional) with payoff matrices, and agent-type modification functions (primitive learning). On the lowest level (Layer-1), we implement BTT (see Fig. 8.3).

In its highest complexity, six parameters (jealousy, power, agreeableness, one signaling type, positive incentive, negative incentive) can be simulated with three different risk calculations of the underlying.

8.4   Results and Findings

We describe results for jealous and egoistic agents for the game "establishing homes." Three types of experimental settings which are exemplary for our investigations are presented.

We started with a finite population of size 100, 1000, and 10,000. We limited our experiments up to 10,000 rounds. From one round to the next, we monitor the different agent populations incorporating the agent characteristics described above. Agent characteristics were modified from one round to the next according to haploid reproduction and simple learning mechanisms by modifying fitness as explained in the previous chapters. In the communication layer of BTT, we apply implicit volatility. We observed in all three settings that after 10,000 rounds and in case of small population sizes even earlier, the population stabilized.

---

[9]In its highest complexity, six parameters (jealousy, power, agreeableness, one signaling type, positive incentive, negative incentive) can be simulated with three risk calculations of the underlying information.

	100 rounds			500 rounds			10 000 rounds		
	egoists	jealous coop.	uncond. coop	egoists	jealous coop.	uncond. coop.	egoists	jealous coop.	uncond. coop.
Population Size									
100	1	29	70	1	0	99	1	0	99
1000	10	439	551	9	389	602	9	0	991
10000	100	4890	5010	100	4681	5219	97	0	9903

Fig. 8.4   Third experiment

First experiment: we assumed 90% unconditional co-operators, 9% jealous co-operators, and 1% egoists as an initial setting. The group fitness at the end was proportional to the group size. The distribution of egoists, jealous co-operators, and unconditional co-operators stayed stable. This is a typical setting for an evolutionary stable strategy (ESS).

Second experiment: we left the number of egoists with 1% but raised the jealous co-operators up to 50%. We observed that the co-operators were extinguished first and after that the jealous co-operators, and at the end, we ended up with egoists only. Finally, the group fitness was much lower than in the first experiment.

Third experiment: same as the second experiment, but we added positive incentives for cooperation in the sense that an agent got additional scores when helping its neighbor building its house. We observed that jealous co-operators migrated to unconditional co-operators. We ended up with slightly less than 1% egoists, and the rest turned into unconditional co-operators. The jealous co-operators were completely extinguished (see Fig. 8.4). The group fitness was proportional to the group size and better than in the first experiment.

Analyzing Fig. 8.4, we observe that there is a convergence even in very large populations. In this case, the system needs more rounds so that a stable state can be reached. Note that stability is not always achievable. However, for the third type of experiment, this was always the case. Therefore, we conclude that not only jealousy but also egoism might be influenced to a certain extent when proper incentives are applied. However, egoists seem to be much more robust toward incentives.

Incentives might have a big influence on jealousy and even on egoists. The results are encouraging since they indicate that jealousy might be suppressed by applying appropriate incentives. The results though seem to contradict Falk and Kosfeld's investigations (Falk & Kosfeld, 2006) who found that the morale of chiefs can be undermined through incentives. In our examples, we experimented with groups without hierarchies. Therefore, both experiments are not directly comparable.

The applicability of this result to human beings remains questionable since jealousy is a complex property of human nature. It might easily happen that a jealous person suppresses its behavior as long as gains from incentives outperform other disadvantages. In the case of absence of incentives, this person might quickly switch back to its old behavior pattern. This might especially be true for the suspicious jealousy type of person.

In this chapter, we have simulated agent behavior based on stochastic interactions. We found similar high variety in the dynamic behavior of agents as in the spatial deterministic approach. The variety depends upon the different specific rules, such as haploid reproduction, or different reinforced coordination strategies and on payoff functions. In any case, it indicates clearly that emotional modeling is dependent upon utility functions and implementation methods. We also did not expect such a high variety in the dynamic behavior of agents in the spatial deterministic approach. Originally, our intention to simulate human emotional behavior with agents was driven by practical limits with human settings. In those settings, humans often cheat, and therefore, the results are questionable. In any case, it can be concluded that the outcome of the simulation of socio-cognitive properties including emotion depends on the applied methodology.

Note that our simulations are based on exogenous parameters and we use game theoretical approaches to analyze social behavior of agents. We are not able to simulate emotions on intra-agent basis. In human brains, emotions and social behavior are driven by bottom-up regulation systems. These systems are driven by neurotransmitters on chemical basis such as dopamine, noradrenalin, or peptides such as endorphins. Some of these transmitters are spread to a couple of neurons in parallel in a diffuse way which is not yet completely understood.

Moreover, even if we could encode all secrets of the human brain, intelligent digital agents simulating all known human properties and emotions simultaneously (digital selves) might cause some intractable problems. Nevertheless, we try to show in the next chapter under which preconditions digital selves might manage digitized service chains.

Part III
A Symbiosis

Here, we present a symbiosis of Parts I and II of this book. We show under which preconditions digital selves might construct and produce digital twins as integrated design elements in transhuman ecosystems. The chapters in this part are dedicated to opportunities and modes of co-creating reflective socio-transhuman systems based on digital twin models, exploring mutual control and continuous development.

The simulation examples of the spatial simulation presented in Part II of this book previously represent examples for Complex Adaptive Systems (CAS) since they show unpredicted emergent patterns in certain parameter regions. Unfortunately, such simulation applications act as stand-alone systems mastering specialized tasks. The big question from the engineering point of view is how to consolidate approaches from Parts I and II.

Structured engineering enables representing digital selves (humanoids) as CAS. They help mastering service and production chains independent from any human intervention. In this respect, System-of-Systems (SoS) design will play a major role since it connects certain systems, such as humanoids, to a sequence of steps utilized by machines which themselves constitute separate systems. Social simulations of collective behavior need to be embedded in a SoS design, so that fully automated service chains can be controlled by humanoids.

We think that subject orientation helps in SoS design and it might therefore contribute to the implementation of intelligent autonomous service chains. In its last consequence, digital selves might construct digital service chains even independent from human intervention. In this respect, digital selves would have to adapt and learn subject-oriented modeling so that they would not only be able to control but also to adapt and design their individual service chains according to their specific needs.

We present the design-integrated engineering approach utilizing subject-oriented modeling and execution and its shortcomings and desirable enhancements for the realization of our visionary goals. By means of a simple example, we discuss issues in designing, modeling, and deploying digital twins representing digital selves in socioeconomic environments.

Chapter 9
System-of-Systems Thinking

Taking into account the continuous generative nature of socio-technical systems, actor intentions and behavior, and, thus, relations may constantly change. We need to delineate their nature, as being transactional, organic, or semi-organic, since they lead to deep changes in the way we act and "produce" and finally affect behavior in and of organizations and societies (Bonfour, 2016). As already discussed in Part I, taking a CAS perspective enables delineation while preserving system thinking in terms of networked but modular elements acting in parallel. In socio-technical settings, these elements can be individuals, technical systems, or their features. As CAS elements, they form and use internal models to anticipate the future, basing current actions on expected outcomes. According to CAS theory, in CAS settings, each element sends and receives signals in parallel, as the setting is constituted by each element's interactions with other elements. Actions are triggered upon other elements' signals. In this way, each element also adapts and, thus, evolves through changes over time.

Self-regulation and self-management have become crucial assets in dynamically changing settings. Self-organization of concerned system elements is considered key in handling requirements for adaptation. However, for self-organization to happen, actors need to have access to relevant information of a situation. Since the behavior of autonomous actors cannot be predicted, some structure is required to guide behavior management according to the understanding of actors and their capabilities to change their situation individually.

From the interaction of the individual system elements arises some kind of global property or pattern, something that could not have been predicted from understanding each particular element. A typical emergent phenomenon is a momentum stemming from an emergency handling actor when deciding upon a certain behavior, such as contacting other actors for specific requests. Global properties result from the aggregate behavior of individual elements.

Applying System-of-Systems (SoS) thinking is considered an effective way of handling CAS, in particular when developing complex artifacts in a structured way (Jamshidi, 2011). According to the Institute of Electrical and Electronics Engineers

© The Author(s) 2022
F. Barachini, C. Stary, *From Digital Twins to Digital Selves and Beyond*,
https://doi.org/10.1007/978-3-030-96412-2_9

(IEEE's) Reliability Society, a system is "a group of interacting elements (or subsystems) having an internal structure which links them into a unified whole. The boundary of a system is to be defined, as well as the nature of the internal structure linking its elements (physical, logical, etc.). Its essential properties are autonomy, coherence, permanence, and organization" (IEEE-Reliability Society Technical Committee on Systems of Systems, 2014).

A System-of-Systems (SoS) is a system that involves several systems "that are operated independently but have to share the same space and somehow cooperate" (*ibid.*, p.2). As such, they have several properties in common: operational and managerial independence, geographical distribution, emergent behavior, evolutionary development, and heterogeneity of constituent systems (ibid.). These properties affect setting the boundaries of SoS and the internal behavior of SoS and, thus, influence methodological SoS developments (Jaradat et al., 2014). SoS are distinct with respect to:

- *Autonomy* where constituent systems within SoS can operate and function independently and the capabilities of the SoS depend on this autonomy
- *Belonging* (integration), which implies that the constituent systems and their parts have the option to integrate to enable SoS capabilities
- *Connectivity* between components and their environment
- *Diversity* (different perspectives and functions)
- *Emergence* (foreseen or unexpected) (*ibid.*)

Several structures and categorization schemes have been used when considering complex systems as System-of-Systems, ranging from close coupling (systems within systems) to loose coupling (assemblage of system). They constitute embodied systems cooperating in an interoperable way (Weichhart et al., 2018), allowing for the autonomous behavior of each system while contributing through collaboration with other systems, in order to achieve the objective of the networked systems (SoS) (Maier, 2014).

Referring to structural and dynamic complexity, structural complexity derives from (1) heterogeneity of components across different technological domains due to increased integration among systems and (2) scale and dimensionality of connectivity through a large number of components (nodes) highly interconnected by dependences and interdependences. Dynamic complexity manifests through the emergence of (unexpected) system behavior in response to changes in the environmental and operational conditions of its components (IEEE-Reliability Society Technical Committee on Systems of Systems, 2014).

A typical technical SoS example is contextualized apps available on a smartphone. Each of them can be considered as a system. When adjusting them along a workflow, for example, to raise alert and guide a patient to the doctor, in case certain thresholds with respect to medical conditions are reached for a specific user, several of these systems, such as the blood pressure app, calendar app, and navigation app, need to be coordinated and aligned for personal healthcare, updating the task manager of the involved users. In this case, the smartphone serves as SoS carrier, supporting the patient-oriented redesign of the workflow and, thus, the SoS structure. The apps of the smartphone can still be used stand-alone, while the smartphone serves as a communication infrastructure and provider of networked healthcare-relevant subsystems. It is the latter property that qualifies the smartphone as a carrier of an SoS.

When we project this concept on understanding complex ecosystem, system elements can become aware of their capability to act autonomously while at the same time being part of a bigger whole, namely, the business organization (or even of several organizations). Awareness of active elements being part of a complex system as a System-of-Systems is considered according to their specific roles in a certain situation. Actors need to become aware of which System-of-Systems they are part of (they can be part of various System-of-Systems). For instance, a digital self can be part of a System-of-Systems consisting of two systems, with one role part of system one and another role in another, larger system.

Chapter 10
Provision of Information as Relational Task

In this section, we discuss a CAS modeling approach of ecosystems targeting the pragmatic aspect when grasping a certain situation through semantic role modeling. The approach enables SoS thinking and will be exemplified by a concrete case from an institutional education. Shchedrovitsky (2014) in his analysis on the engineering nature of organization, leadership, and management of work pinpointed to conveying a specific meaning according to a situation and, thus, grasping situations according to semantics (p. 42 ff):

What is 'meaning'? It is a tricky question. Really, there isn't any meaning. Meaning is a phantom. But here's the trick. I can say a sentence, like 'The clock has fallen off the wall' in two situations with two completely different meanings: 'The clock fell' and 'The clock fell.' The change of accent corresponds to two fundamentally different situations. Imagine this: when I am lecturing, I have got used to the fact that there is a clock here on the wall. At some point, I turn, I see an empty space, and someone in the audience says, 'The clock fell off the wall.' They might simply have said 'it fell' because, in this instance, the word 'clock' carries no new information. I look at the clock, I have got used to it and everyone in the lecture hall has got used to it. We look at that place and someone says 'it fell off the wall', and that phrase provides new information.

But now imagine a different situation. I am giving a lecture and all of a sudden there is a crash behind me. What has made it? I am told, 'The clock fell off the wall.' The situation is entirely different because what is new in this instance is the message about the clock. I heard something fall—that is a given—and I am told that it is the clock that fell. We pin this down in terms of 'subject' and 'predicate' in their functional relationships: in the first case, the clock is the subject, and in the second case the subject is the falling. We carry out syntactical analysis and highlight a difference between the two oppositions 'noun–adjective' and 'subject–predicate'. The distinction between subject and predicate is this: when we have a text, the subject is what we are talking about and the predicate is the characteristic that we ascribe to it. So when I hear any text, I understand it through an analysis: I work out what is the subject. Why do I work it out? I relate it to the situation.

The subject might be an action. In an algorithm I always treat actions as items, to which characteristics are ascribed. So I am always doing a particular sort of work: I parse the text syntactically, identify its syntactical organisation, its predicate structure, and map this onto the situation. This is a process of scanning, of relating the text to the situation. When you understand my text now, you carry out this complex relational work. You are constantly identifying what is being talked about and what I am saying about it. This is the standard

F. Barachini, C. Stary, *From Digital Twins to Digital Selves and Beyond*, https://doi.org/10.1007/978-3-030-96412-2_10

work that goes on automatically, you understand what is being said to the extent that you can find these objects and relate the text to them.

These paragraphs reveal several insights that are not only relevant when trying to capture a situation at hand but also when aiming to represent or modeling it. Hence, providing information needs to be considered a context-dependent process itself. Simply by focusing on a specific part of a sentence, like shown above for "The clock has fallen off the wall," different meanings can be conveyed, and thus, different situations and adjacent work practices could be revealed.

Shchedrovitsky considers ascribing meaning to a situation as relational work. It requires an active entity identifying elements of concern (perceived) information can be assigned to. When we think about selecting a specific meaning out of possible meanings, the initial provision on information for an active system element or actor can have significant influence on the subsequent behavior of this actor and all relations (i.e., the entire system). Hence, in the following, we reflect on modeling the ecosystem in terms of CAS actors using subject-oriented concepts (see Part I of this book) to describe a perceived situation and their use for taking an actor perspective which relates to the semantic and pragmatic quality of models.

We then provide some triggers to rethink how developers elicit and represent situation-sensitive knowledge. A model of eliciting and structuring perceptual knowledge of actors in a certain situation is proposed based on analyzing behavior that could be used as blueprint for engineering digital selves. The model contains several perspectives helping to structure individually perceived situational information for further operation. Each perspective can be enriched with another one leading to a cascade of perspectives, finally allowing to create digital behavior models.

10.1   Basic Model Generation

Entities in terms of networked active elements, termed subjects, act in parallel. Since each of those actors or subjects can be described in terms of its behavior and has the capability to exchange messages, a federated choreographic ecosystem is established:

- Federation means a form or single unit, within which each actor or subject or organization keeps some internal autonomy.

  - This form or single unit identifies the perceived part of the world that is considered relevant to describe a specific situation. It sets up the universe of discourse or context space for representation and action.
  - Keeping some internal autonomy at some point requires to be more concrete: The "some" is dedicated to the level of abstraction considered representative for the stakeholders or modelers, both with respect to functional or technical activities and interaction or communication with other subjects.

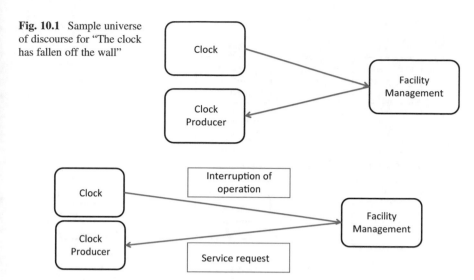

Fig. 10.1 Sample universe of discourse for "The clock has fallen off the wall"

Fig. 10.2 Sample interaction pattern for "The clock has fallen off the wall"

- Choreographic ecosystem refers to recognizing concurrent, however, synchronized processes and activities
  - In a community of interacting elements and their environment
  - When considered as networked or interconnected system

According to this perspective, ecosystems operate as autonomous, concurrent behaviors of distributed subsystems or actors. A subject is a behavioral role assumed by some entity that is capable of performing actions. The entity can be a human, a piece of software, a machine (e.g., a robot), a device (e.g., a sensor), or a combination of these, such as intelligent sensor systems.

Since subjects represent systems with a uniform structure, they can be used to define federated systems or System-of-Systems (SoS), featuring autonomy, coherence, permanence, and organization (cf. IEEE-Reliability Society Technical Committee on Systems of Systems, 2014). SoS subjects can execute local actions that do not involve interacting with other subjects (e.g., a clock providing the time in a classroom) and communicative actions that are concerned with exchanging messages between subjects, i.e., sending and receiving messages, e.g., triggering ringing a tone. Figure 10.1 shows a set of federated systems or subjects, Clock, Facility Management, and Clock Producer, that could be considered of relevance for "The clock has fallen off the wall." The directed links denote the interaction pattern for message exchange.

As already mentioned, each setting or situation can be structured in subject-oriented behavior modeling as a set of individual networked actors or systems, such as facility devices, encoded in subject diagrams according to their communication with each other. Figure 10.2 shows the corresponding for that pattern (cf. Part

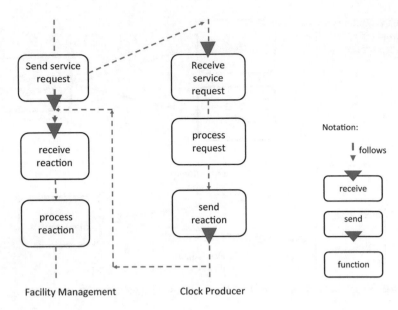

Fig. 10.3 Sample Behavior Synchronization (SBD)

I on behavior modeling). The rectangles denote the messages that the systems exchange.

Figure 10.2 shows a Subject Interaction Diagram (SID). SIDs provide a global view of a SoS, comprising the subjects involved and the messages they exchange. The SID contains a maintenance support process. It comprises several actors (subjects) involved in communication: the Facility Management coordinating all maintenance activities, a Clock Producer taking care of providing a working clock, and the Clock providing scheduling support in classroom management. They exchange messages in case of operational problems as shown along the links between the subjects (rectangles).

Subject Behavior Diagrams (SBDs) provide a local view of the process from the perspective of individual actors (subjects). They include sequences of states representing local actions and communicative actions including sending messages and receiving messages. Arrows represent state transitions, with labels indicating the outcome of the preceding state (see Fig. 10.3). The part shown in the figure represents a service request to the Clock Producer subject from the Facility Management subject.

Given these capabilities, representations are characterized by (1) a simple communication protocol (using SIDs for an overview) and, thus, (2) standardized behavior structures (enabled by send-receive pairs between SBDs), which (3) scale in terms of complexity and scope. They enable CAS behavior pattern specification.

Subject-oriented models are designed to probe representations for operation: once a SBD, e.g., the Facility Management subject, is instantiated, it has to be decided (1) whether a human or a digital device (organizational implementation) and

(2) which actual device is assigned to the subject, acting as technical subject carrier (technical implementation). Typical subjects are devices and their process-specific services, including smartphones, tablets, laptops, healthcare devices, etc. Subjects can also be role carriers controlling or executing tasks. Both types of instantiations can be supported by subject-oriented runtime engines. For a recent overview, see Krenn et al. (2017). These engines provide services linked to some ICT infrastructure.

Once the runtime engine is tightly coupled to model representations, ad hoc and domain-specific requirements can be met dynamically. The situation-sensitive formation of systems and their behavior architecture need to be validated before being executed without further transformation. Hence, models can be adapted according to the SoS their models are part of. In case of existing knowledge based on successful, re-occurring patterns, e.g., for routine tasks, they could be integrated to improve the overall process performance, e.g., including the processing of complex events.

Overall, a subject-oriented representation of any setting can come close to the "reality" as perceived and pictured by humans, both, in terms of its elements as behavioral entities including their set of activities and interactions and in terms of its description as natural language, can directly be used in conveying the content of subject-oriented representations. In order to facilitate practical application (cf. Fleischmann et al., 2015; Neubauer & Stary, 2017), subject-oriented development requires situation-sensitive modeling support from organizational management.

10.2 Capturing Situations

In this section, we sketch behavior modeling for a specific situation, as introduced in Stary (2020b). We detail the methodological approach and exemplify it in a role-specific way.

When actors perceive situations, they start with spotting relevant elements according to their current perspective leading to assigning a specific meaning: ". . . meaning is a particular structural representation—a sort of freezeframe—of the process of understanding. . . . But see what we actually do. Here is a movement. For example, something falls. It leaves a trail. Now we begin to slice this trail into sections, we get parts of the trail and we transfer it to the movement. . . . We divide it into stages and phases, but in order to do this we have to find and register the traces (the trail) of this process" (Shchedrovitsky, 2014, p.43).

The "trail" may range from realizing the trigger event for the clock's falling down to watching how the broken glass spreads over the floor in the glass room. Evaluating this trail allows to scope the entire scene in terms of all relevant elements involved, e.g., the holder went off the wall, the clock fell down, and the clock fell apart when touching the floor. Hence, meaning could be action triggered which in turn is relevant for the stakeholder in the room. Assuming that nobody got hurt through the event, for the students in the room, it may be an event of low complexity, as they do not have to care about the time and are able to watch their steps when avoiding to step in the clock's broken parts. For the teacher, it is a major event, as he/she needs to take care about the time and the safety of the students.

As we can see, each actor constructs meaning through some role-specific glass. It may require immediate action or reaction to an event. The teacher may take action through interrupting the process of teaching and switching to the role of caretaker of classroom safety when warning the student when leaving the classroom. From the teacher's perspective, in a second step, the time problem needs to be addressed, assuming classes are structured along time slots. The teacher needs to interact with somebody from the class or facility management to ensure correct timing, in case he/she relies on an external source of information with respect to time. Finally, the facility management needs to be addressed for taking care of all the damage. From a representational perspective, several entities are involved to make meaning out of a situation:

- The event itself—being an action itself (falling off the wall ending another operation, namely, the time ticking) or "sliced," a set of small actions or events
- The role—student, teacher, caretaker, and facility management
- Actions and interactions, such as teaching and warning the students
- Concerned objects, the clock and the classroom

Each of these elements is constitutional to model representations. Subjects denote roles, encapsulating behavior in terms of doing, sending, and receiving messages. Finally, the concerned objects are addressed in or passed through messages exchanged between subjects.

Conveying meaning to others as another self Situations trigger not only certain behavior but also need to be documented and transferred to others, e.g., to guide further behavior. It could happen that communication is not documented explicitly. Enforcing to think in terms of communication and interaction of actors enables taking further perspectives and further behavior specifications. For instance, the teacher subject (i.e., an actor role) activates the caretaker which in turn activates the facility management.

Aligning selves through goal-oriented behavior abstractions In order to handle a certain situation, it may not be sufficient to develop a chain of interactions from a single perspective. For instance, administration, technically not involved into the clock falling off the wall, needs to be activated to ensure whether the classroom can be utilized by students for the next class. "The work of organisers, leaders and managers has the character of engineering work: it is structural and technical. Organisers, leaders or managers must always be one step ahead; they have to come up with something new" (Shchedrovitsky, 2014, p.7f).

An alignment scheme for individual and organizational activities alignment according to Shchedrovitsky (2014, p.11) is based on actor-specific goals and interaction relations, i.e., to know whom to involve in which way for further operation. As we will see in the following, the goal can help in identifying inten-tional actor performing self-contained tasks according to the perception of a

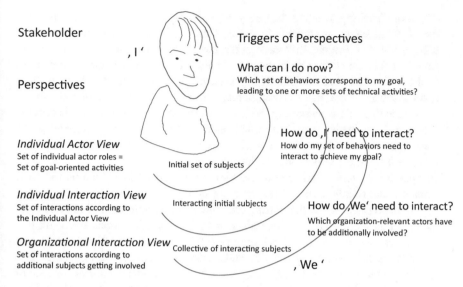

Fig. 10.4 Cascading perspectives

situation. In addition, the means of organizing work could be subject orientation which needs to be probed by applying the model. Perspectives on the situation trigger:

1. *Technical entities* encapsulating behavior by focusing on activities that need to be performed to achieve an objective or implement an intention (usually referring to some task) and, thereby, establishing some functional role
2. *Communication acts* identifying which entity needs to be interacting with another
3. The *mutually adjustment of encapsulated behavior specifications,* as it plays a crucial role not only for acting as a collective in a specific situation but also to complete processes or reach intended goals

Accordingly, the model contains several perspectives helping to structure individually perceived situational information for further operation. Once started with an individual perspective, actors can enrich its result with another one and so on, thus leading to a cascade of perspectives.

Figure 10.4 shows the model serving as a frame of reference of building system capacity based on individually perceived situations. It instantiates Shchedrovitsky's approach in terms of structuring behavior in a goal-oriented way. The left part shows the cascade of perspectives that finally captures the evidence of a specific stakeholder when perceiving and reflecting on a situation:

- *Perspective 1—Individual Actor View*: This perspective captures a set of individual roles in which this stakeholder can act and thinks about in a specific situation. For instance, assuming the clock has fallen off the wall in a classroom with a teacher and students, the teaching role of the teacher addresses all duties related to classroom teaching, whereas the safety-responsibility role of the teacher concerns the physical safety of students in the classroom. Since humans are intentional beings, we can assume that each actor has at least one role or objective to (inter)act that constitutes an actor view. This role or a set of roles corresponds to the individual (task) profile of a person or an artifact. Each role refers to a specific behavior that has a driver, namely, an intention. For instance, the driver of the teaching role is increasing the level of competence of students, whereas the driver of the safety-responsibility role is ensuring the safety of all students in the classroom. Since each role has an intention, each actor can pursue a set of specific goals in a situation, depending on the set of roles.

- *Perspective 2—Individual Interaction View*: This perspective looks on the same situation, but builds upon the results from taking perspective 1 and the further identified roles. It keeps the considered role/objective/intention at the center of interest, but additionally captures a set of individual interactions based on that previously defined intentional behavior set(s). Hence, the set of interactions also depends on the roles in which this stakeholder can act and thinks about in a specific situation. For instance, we assume the stakeholder identifies the role of the teacher (addressing all duties related to classroom teaching) and the safety-responsibility role (ensuring the physical safety of students in the classroom). Then, from this perspective, the stakeholder needs to think about interactions between these two roles. In case the teacher interrupts the class due to the clock's falling off the wall, the safety-responsibility role takes over to ensure the safety of the students in the room. It may lead to ending the class, once the teacher cannot guarantee the safety of the students in this situation, as perceived by this stakeholder. In case the safety-responsibility role does identify safety risks, the safety-responsibility role informs the teaching role to continue teaching. In each case, the stakeholder can provide and specify a set of interactions, for sending and receiving information on a certain topic, involving relevant objects, such as safety measures.

- *Perspective 3—Organizational Interaction View*: This perspective analogously builds upon existing results, this time from taking the previously described perspectives 1 and 2. They already include roles and interactions, however both from an individual perspective. This perspective captures a set of roles this stakeholder perceives to be relevant for a specific situation in addition to the ones he/she can act him-/herself, e.g., taking a community or network perspective. It concerns a set of roles the stakeholder having perspective 1 and 2 cannot take or has no privilege to take. For instance, assuming the clock has fallen off the wall in a classroom with a teacher and students and has been damaging some interior, neither the teaching nor the safety-responsibility role is sufficient to continue with giving a lecture in this classroom. Like from perspective 1, another individual actor view is driven by an intention. In the sample case, the goal could be to keep the classes running that are assigned to this room. Then, the interior needs to be restored, which brings in facility management. Its specific behavior needs to be coupled to the safety-responsibility role, in order to accomplish the respective tasks. Finally, there may be several perspectives related to the "We,"

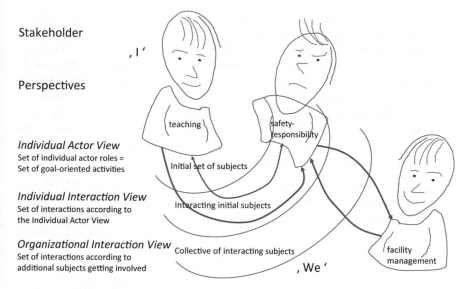

Fig. 10.5   Sample representation

e.g., evolving from an internal community of practice to formal department, networks, regions, and global connections.

Since each perspective builds upon a previous one, a cascade of perspectives evolves in the course of specifying behavior-relevant information. The middle part of Fig. 10.5 reveals the evolving complexity according to refined and networked behavior specifications. The generation of actors and their interaction relations are based on a set of questions that trigger the definition of subjects and their interactions.

- *Initial set of subjects*: The Individual Actor View leads to a set of intentional actor roles that allow stakeholders performing goal-oriented activities. The stakeholder at hand identifies the initial set of behavior abstractions (subjects) by dealing with the question "What can I do now?". This question targets those behavior abstractions that a stakeholder can name, once a goal to be achieved in this situation becomes evident. For instance, in case the clock falls off the wall of the classroom, the ultimate goal of a teacher is to ensure the students' safety before proceeding with the lecture. In order to achieve that goal, the stakeholder can perform a set of technical activities.
- *Interacting initial subjects*: The Individual Interaction View leads to a set of intentional actor roles that synchronize their behavior. The stakeholder at hand identifies all those interactions between the initial set of behavior abstractions (subjects) by dealing with the question "How do 'I' interact?" when having identified more than one role for handling a specific situation. For instance, in case the clock falls off the wall of the classroom, the safety-responsibility role interrupts the teacher to ensure the students' safety before signaling him/her to

proceed with the lecture. Hence, the interactions are defined, in order to achieve the stakeholder goal determined upfront.

• *Collective of interacting subjects*: The Organizational Interaction View leads to a set of intentional actor roles and synchronization of their behavior beyond the stakeholder at hand. This time, he/she needs to answer the question "How do 'We' need to interact?" when embedding further actor roles for handling a specific situation. For instance, in case the clock falls off the wall of the classroom, the safety-responsibility role informs facility management, in case he/she cannot ensure the students' safety. Every interaction with the facility management needs to be defined, in order to achieve the upfront determined stakeholder goal.

Figure 10.5 exemplifies the cascaded perspective. In this case, the stakeholder has identified "teaching" and "safety-responsibility" as role representatives for perspectives 1 and 2 which need to interact sensitive to the safety of the students. For the repair of the clock and classroom restoring, this stakeholder activates facility management through respective interactions.

The "We" perspective can be extended to bring in additional stakeholders, e.g., authorities managing school infrastructures, that are contacted in case needed, e.g., by facility management, to improve the interior. Hence, the number of cascaded perspectives depends on the intention and goal of the stakeholder and results in a systemic view. The schema allows on the one hand focusing on a perceived part of a situation while on the other hand extending perspectives limiting contextual or systemic thinking by enabling interaction links to actor roles valid from other perspectives.

Both elements are essential, as they allow handling complex situations or events without reducing the complexity itself, but rather offering a multi-partite structure. This structure facilitates handling (complex) situations

1. By starting with familiar, since ego-centric behavior encapsulations (roles), and then
2. Stepwise enriching this set of roles by

   (a) Sets of interactions between ego-centric behavior encapsulations
   (b) Including non-familiar behavior encapsulations (roles)
   (c) Coupling them through sets of interaction to all other behavior encapsulations

Hence, without pre-determining the number of perspectives and the number of modeling elements (behavior encapsulations, interactions), a stakeholder can be encouraged to express his/her perception of a situation based on interacting behavior elements. These elements represent subjects allowing to detail pragmatic information in terms of role-specific (internal) behavior. The latter is represented in Subject Behavior Diagrams (SBDs). Given the interaction between the subjects, a SID and, thus, a stakeholder can create a coherent pragmatic model of a situation.

Chapter 11
Enabling Contextual Adaptation

Adaptation during runtime of digital selves in an actor system is enabled through running digital twins as reference subjects. Dynamic adaptation is based on a trigger, such as an emotional signal or the output of a cooperation function, which requires special behavior specification. Message Guards as introduced in Part I of this book allow defining at design time how special context factors influencing behavior can be handled at runtime (i.e., once an actor (subject) has been instantiated).

Switching from routine behavior to another (non-routine) behavior is based on flagging states in the regular behavior sequence serving as triggers and (re-)entry points. In the addressed teaching class example, the Message Guard can be applied when a threshold of students is not willing to leave the room in case an emergency has been reached. Once the flag is raised at runtime, either substitutive procedures that eventually return control to the regular behavior sequence or complementary behavior is triggered that does not return control to the regular behavior sequence.

Message Guards can be flagged in a regular behavior process in any state of an actor when modeled as subjects. The receipt of certain messages, e.g., to abort the process, always results in the same processing pattern. Hence, this pattern should be modeled for every state in which it is relevant. The design decision that has to be taken concerns the way in which the adaptation occurs, either extending an existing behavior or replacing it from a certain state on. In the teaching class example, returning to the original behavior sequence (regular SBD) is given when the emergency case has been handled, and no further intervention of other actors is required when lecturing the class. Replacement of the regular procedure would be required when the SoS Handler subject and, as a follow-up, other actors have to be modified in the behavior sequences.

Message Guards can be applied to any context management issues as part of the digital twin representation. Each digital twin component can be enriched with behavior sequences required for contextual inquiries and processing affecting actor behavior (see Fig. 11.1). The behavior extension is based on the analysis of requirements and messages exchanging context data between the components.

© The Author(s) 2022
F. Barachini, C. Stary, *From Digital Twins to Digital Selves and Beyond*,
https://doi.org/10.1007/978-3-030-96412-2_11

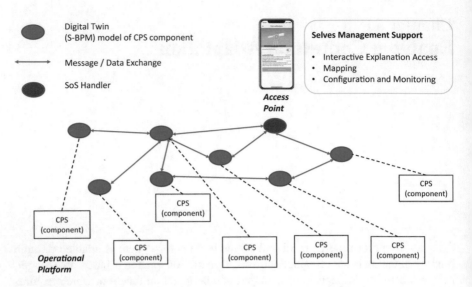

Fig. 11.1 Generic SoS architecture

The component marked in red in Fig. 11.1 is the SoS entry point for handling contextual requirements, e.g., stemming from a certain situation. Each component has a Message Guard for handling SoS requirements, e.g., taking care about the cooperation status of an actor. It communicates with the SoS Handler, e.g., taking into account cooperation with respect to sympathy or other emotional actor states. The latter supports the specification of context requirements and is able to send and receive data from all CPS components.

On request, CPS components deliver data that is matched with context requirements. As the SoS Handler could have access to conventions and regulations, it can recommend activities to actors and address the components involved to meet the contextual requirements. For instance, in case external students should become part of a project team, messages containing information on cooperation willingness are exchanged with other actors until the respective requirements can be met.

Once each actor is represented in an executable subject-oriented model, it can control the exchange of context-relevant data supported by a proper operational platform, such as the compunity suite (http://www.compunity.eu). By monitoring the flow of messages between the networked CPS components, actors can reflect on the current status of achieving contextual requirements. In case they decide to change that flow of information, the approach offers access to the digital twin modeling capability via subject-oriented models. When every twin model can be executed automatically, actors can adapt their requirements to changing context during runtime.

Chapter 12
Embodying Social Behavior

Starting out with human behavior variability in the context of the clock falling off the wall example, we then position algorithmic twin execution for socio-emotional behavior enrichments, utilizing the findings from Part II.

Human behavior is fundamental to human safety and the outcomes of critical situations, such as the clock coming off the wall in a classroom teaching situation. It is correlated with social environments, localization, and other situation-sensitive factors. Once several stakeholders are involved, human-human interactions, i.e., interactions among people or groups of people, and their influence on behavior during critical situations, play a crucial role. They concern interaction with the physical environment, i.e., human-(physical) object interactions, and their handling of critical situations (cf. Zhu et al., 2020). Human-(physical) object interactions involve objects from the environments and their influence on human behavior, such as buildings, including the impact on object performance during critical situations. Handling emergencies concerns human-critical situation interactions, in particular how critical situations impact human behavior and strategies with critical situations. In the following, we apply and adapt the findings by Dovidio (1984) and Zhu et al. (2020) to exemplify the integration of our concepts and modeling approach in Parts I and II.

Critical situations involving physical settings are particularly important, as locations play a key role in people's life, as people tend to spend most of their time in specific locations, such as the workplace, and hence physical settings can be a major factor influencing behavior. For instance, moving along certain routes through the school building can impact how people act in critical situations. They may not be aware that traditionally followed routes need to be reconsidered to avoid disastrous consequences. Due to novel experiences and stress brought about by the critical situation, behavior patterns can emerge that lead to additional fatalities. Thus, a thorough understanding of the relation between human behavior and physical behavior during critical situations is crucial for both critical situation preparedness and response.

© The Author(s) 2022
F. Barachini, C. Stary, *From Digital Twins to Digital Selves and Beyond*,
https://doi.org/10.1007/978-3-030-96412-2_12

Fig. 12.1 Categories of
interactions in critical
situations (Adapted from
Zhu et al., 2020)

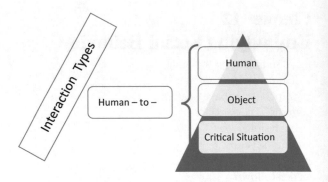

The behavior of an individual during critical situations is interrelated with
(1) other people, e.g., classmates, co-workers, and customers; (2) context attributes,
e.g., location and visibility of objects, including emergency equipment; and (3) attributes referring to the critical situation itself, e.g., broken glass and smoke. Interaction, as reciprocal action or influence of people and/or things on each other, is the
driving force of human behavior and response performance in critical situations. It
can be categorized into human-human, human-(physical) object, and human-critical
situation interactions according to its formation—see Fig. 12.1.

Human-human interactions refer to the collective behavior among people in the
same location and their interactions with people in organization-specific roles, such
as students and facilitators in a classroom, staff members, and critical situation
response teams. Human-object interactions refer to how various physical attributes
(e.g., objects, equipment, exits, or stairs) impact human behavior and how human
behavior (e.g., using familiar walking routes to exit, packing things when leaving)
impacts the local performance during critical situations.

Human-critical situation interactions refer to how critical situations, e.g., presence
of danger in terms of (continuing) damage, fire, and smoke, impact human behavior
and how people cope with criticalities, e.g., removing damaged objects and distinguish fire. The actual behavior is a combination of all interaction types and thus
referred to as second-order behavior. For instance, rushing to the classroom door
encouraging the others to follow while grabbing personal belongings combines all
three types of human interactions.

In our modeling approach, we distinguish the aforementioned categories recognizing the interaction variety—human-human, human-object, human-critical situation, and human-object-critical situation interactions. Since all of these interactions
can play an essential role during critical situations, both for individuals and for group
settings involving different roles, e.g., in facilitators and students in classroom
lessons, we need to detail these interactions to make appropriate decisions related
to behavior design and processes for critical situation handling.

In particular, with respect to human-human interactions among co-located people,
their interactions with people in different roles and the impact of these interactions
on the response performance are of interest. For instance, as previously shown, the
teacher or facilitator in the classroom is engaged in the knowledge transfer (cognitive

behavior) while also being the caretaker on organizational issues, such as starting and closing in time. In the sample case, people are not alone during the critical situation. They are accompanied by others, such as their peers, the facilitator, and those indirectly in the building, including facility managers.

Human-human interactions are one of the most important aspects that determine how people behave and the overall (re-)action behavior and selected behavior patterns. The latter result from people's tendency to observe others' responses to critical situations and select their behavior accordingly. Zhu et al. (2020) could identify several categories of human-human interactions from their study on empirical evidence of behavior: herding, avoiding, grouping, helping and competing, leader-following, and information sharing.

- *Herding behavior* as a specific category of human-human interaction behavior refers to a person following what others are doing, even though the perceived situational information suggests otherwise. For instance, in case of evacuating the classroom, herding behavior refers to a person choosing the most congested route because that route is the most popular choice, instead of alternative routes with less people. Herding behavior can occur when people experience high levels of stress or rationalize as far as they understand the situation. Herding behavior is impacted by both environmental factors, e.g., number of peers, exits nearby, visibility of environment, and personal factors, such as individual attitude.
- *Avoiding behavior* is also related to locality and environmental factors. For instance, in crowded places like lecture halls during lecture time, the level of uncertainty, such as the blocking of visibility through obstacles, is decisive for avoiding the behavior of others. In situations of low uncertainty, e.g., when evacuating classrooms and exits with shorter distance become overcrowded, the majority of people would tend to choose further exits to avoid excessive delays due to heavy congestions.
- *Grouping* is similar to herding and avoiding behavior, but takes into account the connectedness of people. While herding and avoiding behavior may occur among crowds of strangers, grouping behavior requires usually some form of social connectedness. For instance, when people stem from the same peer group, like classmates, they tend to move as a group and look for group members.
- *Helping behavior* is also related to people's pre-existing social as well as emergent collective identities when dealing with critical situations. Thereby, relations among people may be established and might get strengthened through sharing experiences of a critical situation. The latter increases collective identification and, finally, cooperation among people. On the other hand, an increasing level of danger, e.g., when the event of the clock falling off the wall is accompanied by an earthquake, decreases the amount of help.
- *Competing and selfish behavior* can happen when people experience an increased level of stress and loss of personal space. For instance, in case a person feels the need to leave a scene without taking care of others, the behavior could result in competing behavior reaching the exit or another location. The handling of a critical situation for an entire group can be affected by selfish behavior. The

decision on this type of behavior likely depends on pre-existing and/or emerging social relationships, as given in classroom settings.

- *Leader-following behavior* occurs when an occupational and/social role influences people's behavior. Persons can take roles of leaders and followers when critical situations happen. It is based on their knowledge and experience and their personality. It is very common that many people adopt the role of followers during critical situations and less likely that people decide to lead in such situations. Leadership is taken by people with authority who then lead followers according to their individual understanding of the critical situation and the perceived environmental information.

- *Information sharing* is key in critical situations and serves as a medium in human-human interactions. Once a critical event occurs, people start "hunting for information" and make many efforts to grasp more information about the perceived criticality concerning their situation. They start consulting peers or responsible actors, if not forming an ad hoc "crisis committee," to discuss the situation. The information they perceive helps people to evaluate the options when preparing to perform the next activity in a given situation. It can facilitate coping with critical events, but also make handling a critical situation worse for the concerned people. Rather than taking more appropriate actions accordingly, fatalities may be the consequence.

Each of these behavior categories can occur in combination with others and lead to coupled effects. For instance, leader-following and helping behavior can be intertwined, when leaders show helping behavior while guiding people to set specific activities. It may interrupt the leadership function for the sake of completing the support activities. Moreover, persons, when not panicking in critical situations, can process received information, either stemming from the group they have been joining or leading or from the environment.

The environment represents the context of a critical situation. For instance, organizational staff, such as facility managers, can even be first responders to a critical situation, e.g., helping people by leading them to an exit or showing them how to handle emergency equipment. Hence, role-specific behavior not only influences individual behavior but also the relation of interactions, in particular when groups are concerned. Human-object-critical situation interactions play a significant role due to the distinct behavior inputs and patterns of involved stakeholders. For instance, in case facility managers are involved from the beginning in handling the clock case, human-critical situation interaction is influenced by their trained operational behavior. Triggers are trust in role-specific behavior that likely will lead to leader-following patterns.

Figure 12.2 summarizes the mapping of interaction types to model constructs for digital twin representations. Since we follow a communication-centered approach, the representation is based on role or task behavior. Consequently, human-to-object interaction is based on behavioral entities representing individual procedures handling objects or dealing with objects. Since in our case cyber-physical components

Fig. 12.2 Mapping
interaction categories to
modeling capabilities

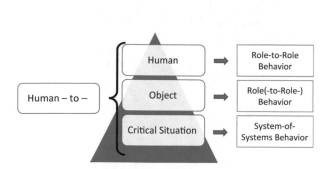

can have dedicated tasks to be accomplished, e.g., pre-processing of sensor data, these interactions can also be of role-to-role type.

Human-to-human interaction is traditionally based on role-to-role behavior. According to the subject-oriented modeling approach, subjects represent role abstractions independent from their actual implementation, and thus, roles can be implemented by physical, digital, or hybrid systems. For human-critical situation interaction, a System-of-Systems mapping is needed, since it requires a baseline, both in terms of what actors are doing and which objects of the environment are concerned in such a specific situation. As such, we consider human-critical situation interaction already as "second-order" interaction, as it implicitly represents human-object-critical situation interaction.

In order to capture the classroom setting, modeling needs to take into account the framing organizational structures (see Chap. 2 in this Part). Students enroll through registration, as shown in the behavior diagrams for course and class registration. Each involved subject, the Student and Course Administration, is modeled by a corresponding behavior diagram (see Fig. 12.3).

Once enrolment has been successful, students attend their classes. The occurrence of a critical situation is captured using Message Guards (1) to decide whether the studying in class can continue as planned or (2) dedicated behavior sequences should be selected for handling the critical situation.

The latter case corresponds to handling a complex event and requires decision-making on behavior options.

Figure 12.4 shows the occurrence of an event from the perspective of a student that leads to a first check whether a critical situation is given or not. This diagram represents human-to-critical situation interactions and implements a System-of-Systems perspective. The Message Guard is activated in case a critical situation is recognized and requires further consideration of the criticality. According to the concept of SoS, the student role triggers the behavior of other actors.

Human-to-object interactions are represented in behavior models as role-to-role interaction when messages report on behavior as a result of object manipulation or location-specific changes. In the first case, data concerning digital or physical objects

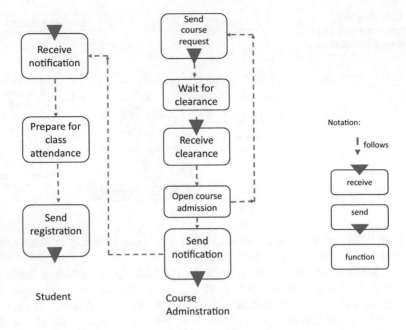

Fig. 12.3 Class registration based on role behavior of the Course Administration and Student

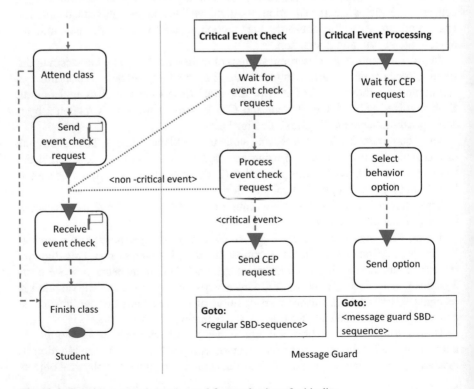

Fig. 12.4 Entering a critical situation and first evaluation of criticality

Fig. 12.5 Adjusting
cooperation behavior based
on fitness of accumulated
knowledge

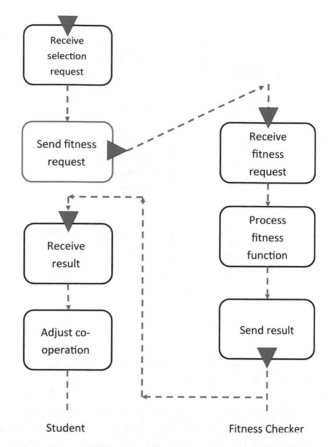

are transmitted via messages to other actors. In the latter case, physical movements
or manipulations of either type are reported to other subjects.

For further modeling, in line with the simulations in Part II of this book, we
assume a population of unconditional co-operators. When we consider the students
as population in which agents can live and work alone or in groups, they can reach a
certain fitness, when they have reached the expected accumulated knowledge (calculated with the business transaction function) that the agent (actor) leaves the
group.

Figure 12.5 shows the activation of the business transaction function for adjusting
the cooperation behavior. The adjustment includes cooperation on a conditional or
unconditional basis in a group. According to the already mentioned types of actors:

- *Jealous co-operators* work conditionally and cooperate until a certain level of
  fitness is reached in their counterparts.
- *Egoists* maximize fitness by working and cooperating only to the extent that the
  expected fitness cost exceeds the expected cost of defecting.
- *Co-operators* work unconditionally and cooperate in any case.

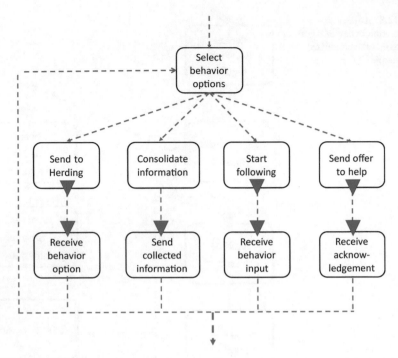

Fig. 12.6 Human-to-human interaction from a student perspective for herding, information sharing, following a leader, and helping

According to these fitness-driven behavior types, (re-)actions in critical situations can be categorized. For instance, the willingness to please others when relating student and facilitator behavior can either be assigned to jealous co-operators or egoists, depending on the fitness threshold to be achieved. Another example is following normative rules, such as waiting for instructions in the classroom, either from facility management outside the room or the facilitator present in the classroom. It can be related to cultural norms including social rewards.

Figure 12.6 exemplifies human-to-human interaction behavior specification when zooming into student behavior after adjusting cooperation behavior based on fitness of accumulated knowledge (see Fig. 12.5). The dotted lines indicated the relative sequence of activities, as there might be additional functions to be performed. However, in this context, only the fundamental patterns of the addressed behavior type are of interest. The examples concern four categories of the aforementioned behavior types:

- *Herding* is further processed to a monitoring subject that reports on all instanti-ated Student subjects during operation. Once behavior changes can be observed, they are reported and received by the student when selecting herding behavior.
- *Sharing information* requires consolidating individual information which is modeled for that case as function state. Once the information has been consoli-dated, it can be broadcasted to all other instances of the Student subject at runtime.

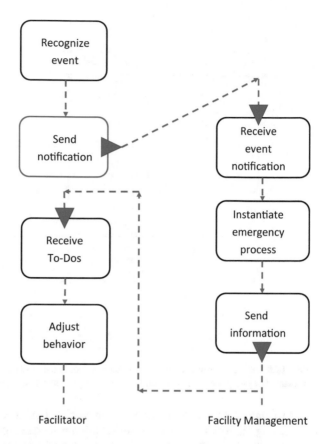

Fig. 12.7 Role-specific behavior of facilitator in critical situation

- *Following* triggers a Leader subject that can be a member of the peer group, i.e., a student, the facilitator, or another role carrier (e.g., facility manager, other community members), that can be addressed at runtime.
- *Helping* is bound to informing others that they can ask for support whenever needed. It is also modeled as broadcast message to instantiated subjects at runtime.

Together with students, a teacher or facilitator is situated in the classroom. Figure 12.7 shows the human-to-human interaction behavior specification for facilitators. Besides adjusting behavior on the fitness of accumulated knowledge, functional interaction with the facility manager is required for further instructions in critical situations. In case of interruption, the facility manager receives a timely report and feeds back the regime to follow.

Figure 12.8 exemplifies some human-to-human interaction zooming into the facilitator behavior specification, including the different types of cooperation behavior on fitness of accumulated knowledge:

- *Egoistic behavior* is encoding when asking for repair, since the facilitator wants to keep a clock in the classroom and asks for prompt replacement.
- *Delegating* expects somebody else, either facility management or even a student to take care about the entire critical situation. It can be trigged by non-cooperative motives.

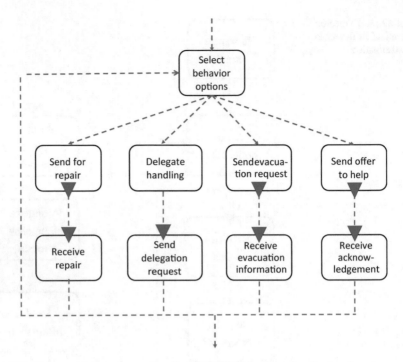

Fig. 12.8 Human-to-human interaction from a facilitator perspective for keeping time control (repair), delegating the further handling, evacuating the room, and helping

- *Evacuating* the room concerns all instantiated subject at runtime and leads to leaving the room according to evacuation principles of the facility. In a way, it corresponds to take leadership making others following.
- *Helping* informs students in the classroom whether they need support handling the critical situation, e.g., when helping to organize another clock or removing broken parts from the classroom. Like the evacuation call, it is modeled as broadcast message to instantiated subjects at runtime.

Now, imagine a combined system. Suppose the teacher is an artificial agent and the crowd consists of humans. If the agent wants to know how the crowd will react when the clock fells down, then the agent asks an oracle. The oracle is a simulation of the crowd, and provided that the individual properties (egoists, conditional co-operators, co-operators) of the humans are known, the oracle will present an answer. Egoists will probably wait until facility management takes over. Conditional jealous people will need some incentives to help. Co-operators will help immediately without waiting for facility management.

In the sense of Daniel Kahneman (2011), the oracle can be seen as the system 2. Similar to people using system 2, when deeper energy-consuming thinking is necessary because pre-learned concepts are not available, machines will use oracles once the machine-learned mental models are not sufficient to solve a problem.

Chapter 13
How to Create Digital Selves

In the following, the construction of digital twin behavior models by means of subject-oriented modeling embedding socio-emotional aspects, and in this way generating digital selves, is considered from a development perspective. We start with providing the fundamental concept before showing the bottom-up and top-down approaches to creating subject-oriented digital selves' representations. Thereby, we structure the adjustment of functional and social modeling for development.

Digital selves form a networked universe of discourse consisting of behavior entities. Hence, their digital twin representations are based on behavior abstractions, also termed subjects. They represent the behavior of any active entity in a socio-technical setting. A specification of a subject does not imply any human, system, or technology that could be used to execute the described behavior. It is a description of setting actions, either in terms of functional operation or communication with other subjects.

Typical examples of subjects are functional roles in organizations and community settings related to some task, such as facilitator, learner, facility manager, back office, and customer service. Less typical behavior abstractions are generic sets of activities, as they refer to cross-organizational or cross-sectoral roles of a domain, however may be critical for operation. Samples of that kind are information collector, privacy management, emergency case handler, or security.

As subjects represent abstractions, any instantiation of a twin model of a digital self needs to be linked to implementation in a socio-technical setting. Sample implementations are robots taking care of facilities in case of emergency, information services provided by concierges, security staff, customer service collaborative robots on how to install IoT sensor systems in a smart home setting, or Enterprise Resource Planning module handling orders.

Digital selves in terms of subjects represent nodes of a network scenario. They communicate with each other by exchanging messages. Typical exchange processes are role-specific subjects, as in a classroom setting a subject "Facilitator" sending the subject "Learner" a learning assignment or in business operation for a subject

© The Author(s) 2022
F. Barachini, C. Stary, *From Digital Twins to Digital Selves and Beyond*,
https://doi.org/10.1007/978-3-030-96412-2_13

"Billing" sending subject "Customer" the billing message. Each message has an identifier in terms of a name and a payload. The name should express the meaning of a message informally. The payloads are the data (business) objects transported along the message exchange.

Digital twin representations as (part of) subject networks encapsulate behavior of entire systems, with each network node or subject characterized by naming internal behavior representations. Internally, each subject either executes local activities ("doing"), or sends messages to other subjects, or expect messages from other subjects ("communicating"). Hence, each subject encapsulates behavior composed of three types of fundamental activities. When recognizing emotional aspects, each activity can be linked to a subject influencing the behavior in subsequent steps. Such "influencer" subjects can be provided with parameters referring to previous activities and deliver decision support for selecting further activities of the calling subject (see also above).

Subjects performing internal activities can be combined in a system behavior specification with separated subject behaviors. Functional samples comprise work tasks, such as handling an order, moderating a discourse between different stakeholders, preparing a decision, or communicating to others. Examples referring to the emotional state of a subject are:

- A subject "Learner" gets uncomfortable with classmates (i.e., other subject carrier of type "Learner") and activates a corresponding pattern of activities, such as questioning the contribution of other subjects "Learner."
- The subject "Facilitator" feels the need to settle an upcoming conflict among subjects of type "Learner" who in turn expect a message "Clarification" from subject "Facilitator."

Communication when based on sending and receiving messages (including data) can lead to patterns of information exchange, including socio-emotional states, which might even lead to overrun institutional regulations requiring formal approval, such as the following example reveals:

- A subject "Facility Manager" feels the need to accelerate evacuation in case of an emergency case and directly requests security guards for help, sending the subject "Safety Unit" directly a message "Clear Building."

Since digital selves and their digital twin representation are embedded in some context, we also need to define that concept for their model representations. The context of each subject is defined by the operating organization and environment it is part of, which in turn is a set of (connected) subjects. Context is also provided by the technological infrastructure by which a subject-oriented behavior specification can be executed.

The System-of-Systems perspective should be implemented, whenever particular concerns, such as handling emergency case, need to be addressed (cf. Heininger & Stary, 2021). It allows to bundle requirements for that concern, e.g., recognizing regulations affecting all actors of the network, and to connect to each actor implementing these requirements. Finally, monitoring and dynamic adaptation of requirements can be performed, and resulting behavior changes can be put to operation by System-of-Systems subjects.

13.1 Representing Case-Sensitive Semantics

As the construction of a digital twin and digital selves by subject-oriented modeling is based on connecting behavioral entities or abstract components involved in a dynamic setting, subjects and their exchanges of messages (interactions) need to be specified in a specific sequence. A complete subject-oriented model requires the specification of the following:

- Concerned universe of discourse, such as a business operation case
- Active components or subjects involved in a specific case or process
- Interactions the active components or subjects are part of
- Messages the active components or subjects send or receive through each interaction
- Behavior of each subject encapsulating functions and interaction activities in terms of sending and receiving messages

Figure 13.1 shows the generic steps how subject-oriented concepts are applied when embedding socio-emotional aspects into functional/cognitive digital representations of digital selves.

Assume the setting is a classroom and the case to be handled is Emergency Management. The subjects that are involved are at least for functional operation Teacher, Student, Facility Management, and Safety Unit. Sample interactions in case of emergency are Teacher-Facility Management and Facility Management-Safety Unit. Thereby, the exchanged messages are Emergency notification, Emergency

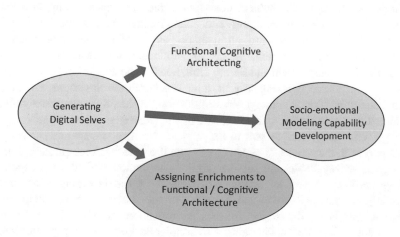

Fig. 13.1 Generating socio-cognitive digital twin representations (digital selves)

confirmation, Evacuation request, and Evacuation plan delivery. The flow of control and information included is as follows: A Teacher sends an Emergency notification to the Facility Management of the building provider. He is going to receive an Emergency confirmation and the Evacuation plan by the Safety Unit.

The corresponding communication structure of that process is represented in a Subject Interaction Diagram. It is the most abstract diagrammatic level of describing behavior encapsulations. Hence, for each active entity, a subject as part of a Subject Interaction Diagram has to be specified. The set of subjects and their message exchanges establish a network as considered universe of discourse. Typically, all relevant actors and components involved in a scenario or operational case are captured by a Subject Interaction Diagram (SID).

At that stage of development, it is not specified whether a subject or role behavior model is implemented by a digital system, a human, or a Cyber-Physical System. For instance, a teacher could either be a human facilitator, a piece of software, or a cyber-physical artifact, including actuators in the classroom. In either case, algorithms that support social behavior decision can be used for decision-making for the next action, in the case of both routine behavior, e.g., supporting learners when accomplishing their assignments, and non-routine behavior, e.g., acting in case of emergency in the classroom. For each activity of a subject detailed in a Subject Behavior Diagram, an algorithm can be activated. This capability enables evaluating socio-emotional behavior patterns, e.g., proposing action in case of an emergency, and thus taking the lead how to proceed in the classroom, when recognizing student herding in a certain direction.

Each behavioral entity (subject) of the SID is refined by its corresponding Subject Behavior Diagram (SBD). A subject's behavior is described by three states (send, receive, internal function) and transitions between these states. Hence, when specifying the behavior of each subject, a sequence of sending and receiving messages and activities to be set for the goal to be achieved or the task that needs to be accomplished need to be represented. The states of a SBD represent operations. They are active elements of the subject description and are typically implemented by services. They enable exchanging and manipulating data or business objects between subjects, including information sharing as part of socio-emotional behavior.

Taking the example of an emergency case, the behavior of each involved subject, such as Facility Management and Safety Unit, needs to be addressed. In the first state of its behavior, the recognizing subject, e.g., Teacher in the classroom, executes the internal function "Prepare emergency notification." When this function is finished, the transition "Notification call" follows. In the succeeding state "send call," the message "emergency call" is sent to the subject "Facility Management." After this message is sent, the subject "Teacher" goes into the state "wait for guidance." If this message is not available, the subject stops its execution until the corresponding message arrives or sets an action independent of the incoming conformation depending on the socio-emotional behavior of the students. Upon receipt of the emergency guidance, the subject follows the transition into state "Wait for safety unit support." In case of changing environmental conditions, e.g., herding of students toward exits instead of waiting for Safety Unit support, the Teacher could take the lead and guide the people to the outside, as captured in the SBD model.

Operations can be of the type "send," "receive," or "internal function": internal functions deal with specific objects, as required in terms of reporting when an emergency case occurs. As a consequence, at least one operation needs to be assigned to each state. Detailing the operations is not necessary at the modeling stage. It is a matter of an abstract object specification or of the integration of an existing component or application. Typical examples of operations are transactions of an Enterprise Resource Planning system processing data, e.g., stemming from updated sensor components when monitoring a classroom situation. Objects are data and/or applications affected by internal functions or operations of a subject and processed through services. For instance, an internal operation "prepare notification" uses internal data to prepare the data for the notification message. This notification data is sent as payload of the message "emergency notification" to its recipient.

13.2 Refining General Behavior Designs

Besides the stepwise refinement of subjects arranged in a Subject Interaction Diagram, an alternative approach to modeling has been developed (see also Fleischmann & Stary, 2012). It starts with an overall generic network model. This generic model represents some kind of chaotic setting: every network element communicates with every other element whenever it wants.

The initial modeling task is therefore to identify the number of elements required for a digital selves' model. This means modelers have to decide upfront how many actors (subjects) are involved in the application case to be described. Starting with a generic network template that is defined by the number of involved actors (subjects), the behavior of a digital selves' ecosystem can become more concrete step by step. The procedure requires several restriction steps:

1. Specify a generic template according to the number of subjects involved in handling a certain operation or business case.
2. Name each subject of the remaining generic template for handling that operation or certain business case, e.g., by identifying its relevant actors.
3. Stepwise reduce the interactions for each subject until the operation or business case can be handled from both a technical (functional) and a socio-emotional perspective.
4. Name each remaining interaction (i.e., message connection) between subjects which are required for handling the business case.
5. Introduce message types according to the functional and socio-emotional modeling needs.
6. Adapt the specification of subject behavior according to the case at hand.
7. Refine the structure of the objects of operation transmitted by the various messages.

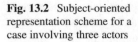

Fig. 13.2 Subject-oriented representation scheme for a case involving three actors

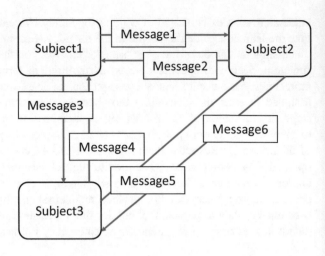

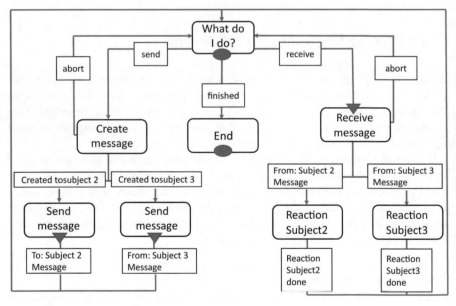

Fig. 13.3 Generic behavior of the subject "subject1"

For the generic model, the question "Who Needs to be Involved?" needs to be answered. For Emergency Management, a generic subject-oriented model with three involved actors (Teacher, Facility Management, Safety Unit) fits to the number of subjects a modeler has to expect for the respective scenarios and related processed (Fig. 13.2). Each actor exchanges messages with another one.

Each subject can send messages with the name "message" to any other subject any time. Figure 13.3 shows the behavior of the subject with the name "subject1." In the select state, a subject decides whether it wants to send or to receive a message. To initiate a system's operation, a subject needs to start sending messages. Choosing the send transition, the subject goes into the state "prepare notification" and provides object data that is transmitted by the message "notification message." After that, the subject decides to which other subject the message with the data as content will be sent.

In the select state, a subject can also decide whether it wants to receive a message1. If there is a message for the subject available, it can be accepted, and a follow-up action can be executed. It is not specified what the follow-up action is. This is like receiving an e-mail. The receiver can interpret the content of an e-mail and knows what the corresponding follow-up action is. The abort transitions back to the select state enable to step back in case a subject has made the improper choice.

Utilizing the message "notification message," a data or business object is sent. The structure of this object corresponds to the structure of a traditional e-mail with extensions like the topic, keywords, and signature.

With each restriction step, the behavior specification is becoming more stringent for the subject holders to their actual situation and way of accomplishing tasks. However, it needs to be noted that bottom-up and top-down modeling not necessarily result in identical models. Nevertheless, both approaches to modeling need to deliver the results from a system behavior perspective.

Addressing the "clock is falling off the wall" case, a facilitator or teacher could be modeled as a digital self by this model. In this case, the behavior is not implemented by a human role taker but some cyber-physical artifact with socio-emotional intelligence. The start state of behavior specification is for both sending and receiving messages "The clock has fallen off the wall." The subsequent figures contain the interaction from both perspectives, sending and receiving information from the facilitator's or teacher's point of view.

The receiver side, as shown in Fig. 13.4, contains receiving hints from students what to do next in the situation that just occurred to the class. In case the facilitator anticipates hints from the student crowd (Kahneman's System 1) (Kahneman, 2011), no further sensing of inputs is performed by the facilitator. The digital self does not set further reactions to received inputs. In case the facilitator does not know the crowd and cannot anticipate the inputs to be received from the students (Kahneman's System 2), a novel type of reaction (solution) needs to be identified, and further analysis of each hint is performed. The development of an action includes simulation of actions and reaction patterns through evaluation before adaptation, e.g., to test opportunistic behavior in terms of common sense reactions to an emergency, such as leaving the classroom to be on the safe side (from various perspectives including safety and role responsibility).

The sender side, as shown in Fig. 13.5, refers to Kahneman's System 2 looking for a socially compatible solution and has its focus on preparing for an acknowledge action by "testing" social appropriateness of possible actions. The "testing" process could be a complete simulation as described in Part II of this book. Depending on the proposed feedback, the Teacher sets the next action.

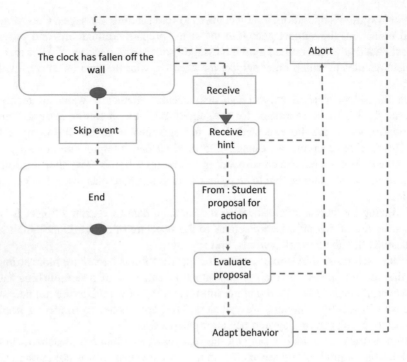

Fig. 13.4 Creating a digital self by restricting behavior of Teacher interacting with Student from a receiver perspective

13.3   The Trans-humanist Development Framework

As demonstrated, digital selves can be modeled by construction or restriction. Socio-emotional consciousness can emerge due to cognitive computing models embodying social behavior models. They lead to digital substrates that either can be grasped in the physical world like Cyber-Physical Systems or remain digital artifacts after being represented as digital twin.

Digital artifacts will increase in everyday life. Hence, we adopted the human-technology-organization framework to show how the individual cognitive perspective on trans-human developments can propagate to digitally enhanced communities.

Figure 13.6 shows the concept we propose for socio-technical development of trans-human settings. The modeling represents the abstraction from biological and technological substrates. The Complex Adaptive Systems perspective recognizes the particular dynamic nature, and the System-of-Systems perspective allows for reducing complexity for the sake of operating and acting in trans-human settings.

For further developments, it seems to be important to exploit the systems engineering skills so that multiple CAS can be modeled in SoS environments properly. The ultimate goal is the incarnation of human-like behaviors as digital selves.

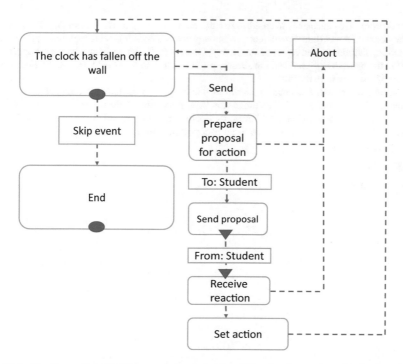

Fig. 13.5 Creating a digital self by restricting behavior of Teacher interacting with Students from the sender perspective

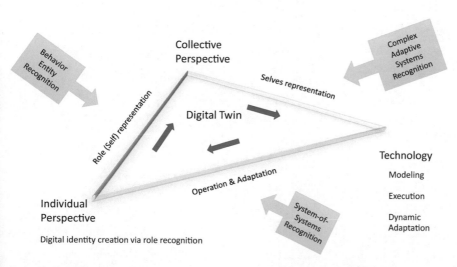

Fig. 13.6 Digital self-development as socio-technical endeavor in the context of behavior representation, Complex Adaptive Systems, and System-of-Systems

Epilogue

Kahneman (2011) describes two different ways the brain forms thoughts. System 1 is the unconscious system which is active most of the time. System 2 is the conscious system which needs much more attention and energy when activated. System 1 represents facts and "learned" mental models. Such models can be digitally presented as facts and rules of a knowledge base. System 2 is activated in our brains when deeper reflections are needed. The digital counterparts for System 2 are cascaded neural networks (deep learning) or simulations. But, as explained in Part II of this book, there is a difference whether the computer simulates a property or whether the machine understands what the simulation is about. We leave it up to the reader to judge whether a digital system might achieve consciousness or whether consciousness is only achievable with biological systems. However, this question might be irrelevant for further developments.

More relevant seems to be the question how humans can coexist with digital selves. Although today humans understand the underlying computing methods, in some cases they cannot explain why AI programs already produce better results in certain domains. In the distant future, the interpretation of big data might equip digital selves with an intelligence which deviates increasingly from human intelligence. This is due to the fact that a digital System 2 is capable of accessing more and more data, whereas the human brain capacity is limited. In its last consequence, humans will neither understand the underlying methods nor the results of the actions performed by digital selves. This poses some risks for humankind.

For a peaceful and frictionless coexistence, it seems to be crucial that digital selves can "understand" human behavior and reactions. Therefore, digital selves need to "understand" ethics, morale, and emotions of human beings. It seems to be crucial that humanity-inspired AI is implemented in our proposed human-technology-organization framework. Thus, the incarnation of human-like behaviors as digital selves seems to be mandatory. Vice versa, it will be important that digital selves can explain their actions and decisions to humans. This is an even bigger challenge since digital selves might develop complete other personalities which might enslave humans.

© The Author(s) 2022
F. Barachini, C. Stary, *From Digital Twins to Digital Selves and Beyond*,
https://doi.org/10.1007/978-3-030-96412-2

Anyhow, digital selves need electricity like humans need food. In its last conse-quence, humans can unplug the socket and thus gain full control again. But what if the transformation of humans to digital substrates will already be too far elaborated at that time? If such a tipping point is exceeded, it may cause serious control problems.

Picture our traffic system with both fully autonomous cars and human-driven cars as socially equal components. In case autonomous cars have some socio-emotional particularities in contrast to human-driven ones, their socio-emotional behavior may have significant impact like human behavior has today. Fully autonomous cars may sense a variety of data and develop patterns of action or reaction that could become relevant for driving. Consider aggressiveness. The sooner one party in traffic anticipates or recognizes aggression, the better behavior could be adapted and conflict could be resolved in novel ways. Collective learning loops as instantiations of System 2 elements could help consolidate socio-emotional intelligence in specific ecosystems. Social norms could even surpass juridical regulations and organiza-tional norms due to their high rate of acceptance.

Whether a complete digital transformation of humans is achievable in due time depends on the survival of humankind and on technologies yet to be invented. Some dystopian thoughts might be in vain if global warming, starvation, a pandemic crisis, atomic wars, an extraterrestrial impact, or simply human stupidity cannot be stopped. In this case, our species will be extinguished before it can transform to computational substrates—*homo homini lupus.*

After all, we should not overestimate AI. Computational substrates are only one possible vehicle to extend life. Bacteria are able to survive many hundred degrees near volcanic flows. Biological computing is right at the beginning like quantum computing was 10 years ago. Maybe we underestimate the ability of the evolutional adaption of organic cells to environmental change. In any case, *nothing* lasts for *eternity.*

References

Ackerman, M., Goggins, S., Herrmann, T., Prilla, M., & Stary, C. (2018). *Designing healthcare that works: A sociotechnical approach*. Academic Press/Elsevier.

Allee, V. (2003). *The future of knowledge: Increasing prosperity through value networks*. Butterworth-Heinemann.

Allee, V. (2009). Value-creating networks: Organizational issues and challenges. *The Learning Organization, 16*(6), 427–442.

Amir, E., Anderson, M. L., & Chaudhri, V. K. (2007). *Report on DARPA Workshop on Self Aware Computer Systems*. SRI International.

Andersen, R. E., Madsen, S., Barlo, A. B., Johansen, S. B., Nør, M., Andersen, R. S., & Bøgh, S. (2019). Self-learning processes in smart factories: Deep reinforcement learning for process control of robot brine injection. *Procedia Manufacturing, 38*, 171–177.

Akerlof, G. A. (1970). The market for lemons: Quality uncertainty and the market mechanism. *Quarterly Journal of Economics, 84*, 488–500.

Axelrod, R. (1984). *The evolution of cooperation*. Basic Books.

Axelrod, R. (1986). An evolutionary approach to norms. *American Political Science Review, 80*(4), 1095–1111.

Axelrod, R. (1997). Advancing the art of simulation in the social sciences. In *Simulating social phenomena* (pp. 21–40). Springer.

Barr, A., Feigenbaum, E. A., & Cohen, P. R. (Eds.). (1981). *The handbook of artificial intelligence* (Vol. 1). William Kaufmann.

Baar, B. J. (1988). *A cognitive theory of consciousness*. Cambridge University Press.

Barachini, F. (1991). The evolution of PAMELA. *Expert Systems the International Journal of Knowledge Engineering, 8*(2), 87–97.

Barachini, F. (2009). Cultural and social issues for knowledge sharing. *Journal of Knowledge Management, 13*(1), 98–110.

Barachini, F. (2007). The business transaction theory and moral hazards for knowledge sharing. In S. Hawamdeh (Ed.), *Creating collaborative advantage through knowledge and innovation* (Series on innovation and knowledge management) (Vol. 5, pp. 1–13). World Scientific.

Barachini, F. (2015). The role of power and jealousy in cooperation—A framework for artificial agents. *Journal of Information & Knowledge Management, 14*(01), 1550009.

Baum, S. D., Armstrong, S., Ekenstedt, T., Häggström, O., Hanson, R., Kuhlemann, K., Maas, M. M., Miller, J. D., Salmela, M., Sandberg, A., Sotala, K., Torres, P., Turchin, A., & Yampolskiy, R. V. (2019). Long-term trajectories of human civilization. *Economics*. Faculty Publications, Smith College, Northampton, MA. https://scholarworks.smith.edu/eco_facpubs/27.

© The Author(s) 2022
F. Barachini, C. Stary, *From Digital Twins to Digital Selves and Beyond*,
https://doi.org/10.1007/978-3-030-96412-2

Bertino, E., Choo, K. K. R., Georgakopolous, D., & Nepal, S. (2016). Internet of things (IoT): Smart and secure service delivery. *ACM Transactions on Internet Technology (TOIT), 16*(4), 22.

Bhatt, C., Dey, N., & Ashour, A. S. (Eds.). (2017). *Internet of things and big data technologies for next generation healthcare.* Springer.

Bishop, J. M. (2018). Is anyone home? A way to find out if AI has become self-aware. *Frontiers in Robotics and AI, 5,* 17.

Bonfour, A. (2016). *Digital futures, digital transformation from lean production to acceluction.* Springer.

Bostrom, N. (2009). The future of humanity. In J.-K. B. Olsen, E. Selinger, & S. Riis (Eds.), *New waves in philosophy of technology* (pp. 186–216). Palgrave McMillan.

Bostrom, N. (2014). *Superintelligence: Paths, dangers, strategies.* Oxford University Press.

Boole, G. (1854, 2016). *An investigation of the Laws of Thought, on which are founded the mathematical theories of Logic and probabilities.* Walton and Maberly, IP Publications.

Bowles, S. (2008). Policies designed for self-interested citizens may undermine "the moral sentiments": Evidence from economic experiments. *Science, 320*(5883), 1605–1609.

Brockner, J. (1988). *Self-esteem at work: Theory, research, and practice.* Lexington Books.

Can, W. S. R., & Seibt, S. D. J. (2016). Making place for social norms in the design of human-robot interaction. In *What social robots can and should do: Proceedings of Robophilosophy/ TRANSOR* (pp. 290–303).

Chen, M., Li, W., Sun, S., Wang, J., & Xia, C. (2016). Evolution of cooperation in the spatial public goods game with adaptive reputation assortment. *Physics Letters A, 380*(1–2), 40–47.

Castelfranci, C. (2000). Affective appraisal vs. cognitive evaluation in social emotions and interactions. In *Affective interactions. Towards a new generation of computer interfaces* (pp. 76–101). Springer.

Castelfranchi, C. (2006). Cognitive architecture and contents for social structures and interactions. In *Cognition and multi-agent interaction* (pp. 355–390). Cambridge University Press.

Chan, S. (2001). *Complex adaptive systems.* ESD 83 Research Seminar in Engineering Systems, 31.

Church, A. (1941). *The calculi of lambda-conversion.* Princeton University Press.

Conte, R., Castelfranchi, C., & Pedone, R. (1999). The impossibility of modelling cooperation in PD-game. In C. Meyer & P.-Y. Schobbens (Eds.), *Formal methods of agents* (Vol. LNAI 1760, pp. 74–89). Springer.

Conte, R., & Castelfranchi, C. (1995). Understanding the functions of norms in social groups. In *Artificial societies; The computer simulation of social life* (pp. 352–267). UCL Press.

Carley, K., & Newell, A. (1994). The nature of social agent. *Journal of Mathematical Sociology, 19*(4), 221–262.

Chong, S. Y., Humble, J., Kendall, G., Li, J., & Yao, X. (2007). Iterated prisoner's dilemma and evolutionary game theory. In *The iterated prisoners' dilemma: 20 years on* (pp. 23–62).

Dehaene, S., & Changeux, J. P. (2011). Experimental and theoretical approaches to conscious processing. *Neuron, 70*(2), 222–227.

Dovidio, J. F. (1984). Helping behavior and altruism: An empirical and conceptual overview. *Advances in Experimental Social Psychology, 17,* 361–427.

Durão, L. F. C., Haag, S., Anderl, R., Schützer, K., & Zancul, E. (2018). Digital twin requirements in the context of industry 4.0. In *Proceedings of the IFIP international conference on product lifecycle management* (pp. 204–214). Springer.

Elster, J. (1996). Rationality and the emotions. *The Economic Journal, 106*(438), 1386–1397.

Elstermann, M., & Ovtcharova, J. (2018). Sisi in the ALPS: A simple simulation and verification approach for PASS. In *Proceedings of the 10th International Conference on Subject-Oriented Business Process Management* (pp. 1–9). Springer.

Falk, A., & Kosfeld, M. (2006). The hidden costs of control. *American Economic Review, 96*(5), 1611–1630.

Fehr, E., & Rockenbach, B. (2003). Detrimental effects of sanctions on human altruism. *Nature, 422,* 137–140.

Fichter, L. S., Pyle, E. J., & Whitmeyer, S. J. (2010). Expanding evolutionary theory beyond Darwinism with elaborating, self-organizing, and fractionating complex evolutionary systems. *Journal of Geoscience Education, 58*(2), 58–64.

Firestone, J. M., & McElroy, M. W. (2003). *Key issues in the new knowledge management.* Routledge.

Fleischmann, A., Schmidt, W., Stary, C., Obermeier, S., & Börger, E. (2012a). *Subject-oriented business process management.* Springer Science & Business Media.

Fleischmann, A., Schmidt, W., Stary, C., & Strecker, F. (2012b). Nondeterministic events in business processes. In *International Conference on Business Process Management* (pp. 364–377). Springer.

Fleischmann, A., & Stary, C. (2012). Whom to talk to? A stakeholder perspective on business process development. *Universal Access in the Information Society, 11*, 125–150.

Fleischmann, A., Schmidt, W., & Stary, C. (Eds.). (2015). *S-BPM in the wild: Practical value creation.* Springer.

Frijda, N. H. (1986). *The emotions.* Cambridge University Press.

Gilchrist, A. (2016). *Industry 4.0: The industrial internet of things.* Apress.

Gross, T., Stary, C., & Totter, A. (2005). User-centered awareness in computer-supported cooperative work-systems: Structured embedding of findings from social sciences. *International Journal of Human-Computer Interaction, 18*(3), 323–360.

Giloi, W. (1997). Konrad Zuse's Plankalkül: The first high-level "non von Neumann" programming language. *IEEE Annals of the History of Computing, 19*(2), 17–24.

Glickstein, M. (1988). The discovery of the visual cortex. *Scientific American, 259*(3), 84–91.

Goldblatt, M. (2002). DARPA's programs in enhancing human performance. In: M. C. Roco, & J. Schummer (Eds.), *Converging technologies for improving human performance: Nanotechnology, biotechnology, information technology and the cognitive science* (pp. 337–341). NBIC-report. National Science Foundation.

Gonzalez-Jimenez, H. (2018). Taking the fiction out of science fiction: (Self-aware) robots and what they mean for society, retailers and marketers. *Futures, 98*, 49–56.

Gödel, K. (1931). Über formal unentscheidbare Sätze der Principia Mathematica und verwandter Systeme I. *Monatshefte für Mathematik und Physik, 38*, 173–198.

Gratch, J., Mao, W., & Marsella, S. (2006). Modeling social emotions and social attributions. In *Cognition and multi-agent interaction* (pp. 219–251). Cambridge University Press.

Grieves, M. (2014). Digital Twin: Manufacturing excellence through virtual factory replication: A whitepaper. Accessed November 3, 2021, from http://innovate.fit.edu/plm/documents/doc_mgr/912/1411.0_ Digital_Twin_White_Paper_Dr_Grieves.pdf.

Grieves, M. W. (2019). Virtually intelligent product systems: Digital and physical twins, Ch. 7. In S. Flumerfelt, K. G. Schwartz, D. Mavris, S. Briceno (Eds.), *Complex systems engineering: Theory and practice* (pp. 175–200). American Institute of Aeronautics and Astronautics.

Guo, B., Zhang, D., & Wang, Z. (2011). Living with internet of things: The emergence of embedded intelligence. In *International Conference on Internet of Things* (pp. 297–304). IEEE.

Guo, B., Zhang, D., Wang, Z., Yu, Z., & Zhou, X. (2013). Opportunistic IoT: Exploring the harmonious interaction between human and the internet of things. *Journal of Network and Computer Applications, 36*(6), 1531–1539.

Hamilton, W. D. (1964a). The genetical evolution of social behaviour I. *Journal of Theoretical Biology, 7*, 1–16.

Hamilton, W. D. (1964b). The genetical evolution of social behaviour II. *Journal of Theoretical Biology, 7*, 17–52.

Hamilton, W. D. (1996). *Narrow roads of gene land: Vol. 1: Evolution of social behaviour.* Oxford University Press on Demand.

Hamilton, W. D. (2001). *Narrow roads of gene land: Vol. 2: Evolution of sex.* Oxford University Press on Demand.

Harari, Y. N. (2019). *A brief history of humankind.* Random House.

Haraway, D. (2015). Anthropocene, capitalocene, plantationocene, chthulucene: Making kin. *Environmental Humanities, 6*(1), 159–165.

Hebb, D. O. (1949). *Organization of behavior*. Psychology Press Edition.

Heininger, R., & Stary, C. (2021). Capturing autonomy in its multiple facets: A Digital Twin approach. In *Proceedings of the 2021 ACM Workshop on Secure and Trustworthy Cyber-Physical Systems* (pp. 3–12).

Hofbauer, J., & Sigmund, K. (2003). Evolutionary game dynamics. *Bulletin of the American Mathematical Society, 40*(4), 479–519.

Holland, J. H. (1992). Complex adaptive systems. *Daedalus, 121*(1), 17–30.

Holland, J. H. (2006). Studying complex adaptive systems. *Journal of Systems Science and Complexity, 19*(1), 1–8.

Hopfield, J. (1982). Neural Networks and physical systems with emergent collective computational abilities. *Proceedings of the National Academy of Science, 79*, 2554–2558.

Hopfield, J. (1984). Neurons with graded response have collective computational properties like those of two-state neurons. *Proceedings of the National Academy of Science, 81*, 3088–3092.

Hopfield, J., & Tank, D. (1985). Neural computations of decisions in optimization problems. *Biological Cybernetics, 52*, 141–152.

Hopfield, J., & Tank, D. (1986). Computing with neural circuits. *Science, 233*, 625–633.

IEEE-Reliability Society Technical Committee on 'Systems of Systems'. (2014). *Systems*. White Paper, 5 p. IEEE, IEEE Society Press.

Jaradat, R. M., Keating, C. B., & Bradley, J. M. (2014). A histogram analysis for system of systems. *International Journal of System of Systems Engineering, 5*(3), 193–227.

Jamshidi, M. (Ed.). (2011). *System of systems engineering: Innovations for the twenty-first century* (Vol. 58). Wiley.

Jaspers, K. (2010). *The origin and goal of history*. Routledge.

Jones, D., Snider, C., Nassehi, A., Yon, J., & Hicks, B. (2020). Characterising the Digital Twin: A systematic literature review. *CIRP Journal of Manufacturing Science and Technology, 29*, 36–52.

Kahneman, D. (2011). *Thinking, fast and slow*. Macmillan.

Kandel, E. R. (2018). *The disordered mind: What unusual brains tell us about ourselves*. Hachette UK.

Kenney, M., & Haraway, D. (2015). Anthropocene, capitalocene, plantationocene, chthulucene; Donna Haraway in conversation with Martha Kenney. In H. Davis & E. Turpin (Eds.), *Art in the Anthropocene: Encounters among aesthetics, environments and epistemologies* (pp. 229–244). Open Humanity Press.

Kidd, C. (2019). *What is the Internet-of-Behavior. IoB explained*. The Business of IT Blog, BMC. Accessed February 10, 2020, from https://www.bmc.com/blogs/iob-internet-of-behavior/.

Kohonen, T. (1982). Self-organized formation of topologically correct feature maps. *Biological Cybernetics, 43*(1), 59–69.

Kohonen, T. (1984). *Self-organization and associative memory*. Springer.

Kohonen, T., Mäkisara, K., Simula, O., & Kangas, J. (1991). *Artificial neural networks*. North-Holland.

Krenn, F., & Stary, C. (2016). Exploring the potential of dynamic perspective taking on business processes. *Complex Systems Informatics and Modeling Quarterly, 8*, 15–27.

Krenn, F., Stary, C., & Wachholder, D. (2017). Stakeholder-centered process implementation: Assessing S-BPM tool support. In *Proceedings of the 9th Conference on Subject-Oriented Business Process Management* (pp. 1–11). Springer.

Kwon, D., Hodkiewicz, M. R., Fan, J., Shibutani, T., & Pecht, M. G. (2016). IoT-based prognostics and systems health management for industrial applications. *IEEE Access, 4*, 3659–3670.

Kurzweil, R. (2005). *The singularity is near: When humans transcend biology*. Viking.

Kurzweil, J. (2012). *Generalized ordinary differential equations: Not absolutely continuous solutions* (Vol. 11). World Scientific.

Kurzweil, R. (2014). *How to create a mind*. Duckworth, an imprint of Prelude Books.

Küster, D., Putze, F., Alves-Oliveira, P., Paetzel, M., & Schultz, T. (2020). Modeling socio-emotional and cognitive processes from multimodal data in the wild. In *Proceedings of the 2020 International Conference on Multimodal Interaction* (pp. 883–885). ACM.

Laird, J. E. (2012). *The soar cognitive architecture*. MIT Press.

Laird, J. E., Newell, A., & Rosenbloom, P. S. (1987). SOAR: An architecture for general intelligence. *Artificial Intelligence, 33*(3), 1–64.

Larivière, B., Bowen, D., Andreassen, T. W., Kunz, W., Sirianni, N. J., Voss, C., Wünderlich, N. V., & De Keyser, A. (2017). "Service Encounter 2.0": An investigation into the roles of technology, employees and customers. *Journal of Business Research, 79*, 238–246.

Lee, E. A. (2008). Cyber physical systems: Design challenges. In *11th IEEE International Symposium on Object and Component-Oriented Real-Time Distributed Computing* (pp. 363–369).

Lazarus, R. (1991). *Emotion and adaption*. Oxford University Press.

Li, S., Da Xu, L., & Zhao, S. (2018). 5G Internet of Things: A survey. *Journal of Industrial Information Integration, 10*, 1–9.

Li, J., & Kendall, G. (2014). The effect of memory size on the evolutionary stability of strategies in iterated prisoner's dilemma. *IEEE Transactions on Evolutionary Computation, 18*(6), 819–826.

Libet, B. (2005). *Mind time: The temporal factor in consciousness*. Harvard University Press.

Löwenstein, G. (1996). Out of control: Visceral influences on behavior. *Organizational Behavior and Human Decision Processes, 65*(3), 272–292.

Lu, W., Wang, J., & Xia, C. (2018). Role of memory effect in the evolution of cooperation based on spatial prisoner's dilemma. *Physics Letters A, 382*(42–34), 3058–3063.

Macy, M. W. (1998). Social order in artificial worlds. *Journal of Artificial Societies and Social Simulation, 1*(1), 4.

McCarthy, J. (2004). *Notes on self-awareness*. University of Stanford, Research Report.

Maier, M. W. (2014). The role of modeling and simulation in system of systems development. In *Modeling and simulation support for system of systems engineering applications* (pp. 11–41).

Moore, G. E. (1965). Cramming more components onto integrated circuits. *Electronics, 38*(8).

More, M. (1990). Transhumanism: Towards a futurist philosophy. *Extropy, 1990*(6), 6–12.

More, M., & Vita-More, N. (Eds.). (2013). *The transhumanist reader: Classical and contemporary essays on the science, technology, and philosophy of the human future*. Wiley.

Maynard-Smith, J., & Price, G. R. (1973). The logic of animal conflict. *Nature, 246*, 15–18.

Maynard-Smith, J. M. (1982). *Evolution and the theory of games*. Cambridge University Press.

Nash, J. (1950). Equilibrium points in N-person games. *Proceedings of the National Academy of Science, 36*, 48–49.

Neubauer, M., & Stary, C. (2017). *S-BPM in the production industry: A stakeholder approach* (p. 232). Springer.

Newell, A., & Simon, H. A. (1972). *Human problem solving*. Prentice Hall.

Ohtsuki, H., & Iwasa, Y. (2004). How should we define goodness? Reputation dynamics in indirect reciprocity. *Journal of Theoretical Biology, 231*, 107–120.

Oppl, S., & Stary, C. (2019). *Designing digital work: Concepts and methods for human-centered digitization*. Springer Nature.

Paiva, A. (2020). *Affective interactions: Towards a new generation of computer interfaces*. Springer.

Panetta, K. (2019). Gartner top strategic predictions for 2020 and beyond. Accessed February 10, 2020, from https://www.gartner.com/smarterwithgartner/gartner-top-strategic-predictions-for-2020-and-beyond/.

Pelham, B. W., & Swann, W. B., Jr. (1989). From self-conceptions to self-worth: On the sources and structure of global self-esteem. *Journal of Personality and Social Psychology, 57*, 672–680.

Pang, T. Y., Pelaez Restrepo, J. D., Cheng, C. T., Yasin, A., Lim, H., & Miletic, M. (2021). Developing a digital twin and digital thread framework for an 'Industry 4.0' shipyard. *Applied Sciences, 11*(3), 1097.

Pera, R., Occhiocupo, N., & Clarke, J. (2016). Motives and resources for value co-creation in a multi-stakeholder ecosystem: A managerial perspective. *Journal of Business Research, 69*(10), 4033–4041.

Picard, R. W. (1997). *Affective computing*. MIT Press.

Rapoport, A. (1999). *Two-person game theory*. Dover.

Rosly, M. A., Miskam, M. A., Shamsuddin, S., Yussof, H., & Zahari, N. I. (2020). *Data linking testing between humanoid robot and IoRT network server for autism telerehabilitation system development* (pp. 161–169). RITA 2018, Springer.

Rosenblatt, F. (1958). The perceptron. A probabilistic model for information storage and organization in the brain. *Psychological Reviews, 65*, 386–408.

Rosenberg, M. (1965). *Society and the adolescent self-image*. Princeton University Press.

Rydell, R. J., & Bringle, R. G. (2007). Differentiating reactive and suspicious jealousy. *Social Behavior and Personality, 35*(8), 1099–1114.

Sadoughi, F., Behmanesh, A., & Sayfouri, N. (2020). Internet of things in medicine: A systematic mapping study. *Journal of Biomedical Informatics, 103*, 103383.

Savitha, J., & Akhilesh, K. B. (2020). Conceptualizing the potential role of IoT-enabled monitoring system in deterring counterproductive work behavior. In K. Akhilesh & D. Möller (Eds.), *Smart technologies. Scope and applications* (pp. 111–120). Springer.

Schmidt, M. P. (2000). *Knowledge communities: Mit virtuellen Wissensmärkten das Wissen in Unternehmen effektiv nutzen*. Addison-Wesley (Imprint of Pearson Education Germany).

Sharpe, F. W., Alexander, G. J., & Bailey, V. J. (1995). *Investments*. Prentice Hall.

Shannon, C. (1938). A symbolic analysis of relay and switching circuits. *Transactions of the American Institute of Electrical Engineers, 57*(12), 713–723.

Shchedrovitsky, G. P. (2014). Part I Selected Works: A guide to the methodology of organisation, leadership and management. In V. B. Khristenko, A. G. Reus, A. P. Zinchenko, et al. (Eds.), *Methodological school of management*. Bloomsbury.

Sigmund, K., Hauert, C., & Nowak, M. A. (2001). Reward and punishment. *Proceedings of the National Academy of Science USA, 98*(19), 10757–10762.

Searle, J. R. (1980). Minds, brains, and programs. *Behavioral and Brain Sciences, 3*(3), 417–424.

Sodero, A., Jin, Y. H., & Barratt, M. (2019). The social process of Big Data and predictive analytics use for logistics and supply chain management. *International Journal of Physical Distribution & Logistics Management, 49*(7), 706–726. https://doi.org/10.1108/IJPDLM-01-2018-0041

Spence, A. M. (1973). Job market signaling. *Quarterly Journal of Economics, 87*, 355–374.

Spindler, M., & Stary, C. (2019). Co-vival: Embracing artificial and human intelligences an awareness approach for transhuman futures. *Challenging Organizations and Societies. ® reflective hybrids, 8*(1), 1303–1359.

Staller, A., & Petta, P. (2001). Introducing emotions into the computational study of social norms: A first evaluation. *Journal of Artificial Societies and Social Simulation, 4*(1).

Stary, C. (2014). Non-disruptive knowledge and business processing in knowledge life cycles – Aligning value network analysis to process management. *Journal of Knowledge Management, 18*(4), 651–686.

Stary, C., & Fuchs-Kittowski, K. (2020). Zur Wiedergewinnung des Realismus als notwendige Grundlage einer am Menschen orientierten Informationssystemgestaltung und Softwareentwicklung. *Leibniz Online 41*, Zeitschrift der Leibniz-Sozietät der Wissenschaften zu Berlin.

Stary, C. (2020a). Co-creation in transhuman realities: Setting the stage for transformative learning. In *Transhumanism: The proper guide to a posthuman condition or a dangerous idea?* (pp. 225–244). Springer.

Stary, C. (2020b). The Internet-of-Behavior as organizational transformation space with choreographic intelligence. In *Proceedings of the International Conference on Subject-Oriented Business Process Management* (pp. 113–132). Springer.

Stary, C. (2021). Digital Twin generation: Re-conceptualizing agent systems for behavior-centered cyber-physical system development. *Sensors, 21*(4), 1096.

Stiglitz, J. (1975). The theory of screening, education and the distribution of income. *American Economic Review, 65*, 283–300.

Sun, R. (2006). *Cognition and multi-agent interaction.* Cambridge University Press.

Sun, R. (2003). *A tutorial on CLARION 5.0.* http://www.cogsci.rpi.edu/rsun/sun.tutorial.pdf.

Tan, V., & Varghese, S. A. (2016). IoT-enabled health promotion. In *Proceedings of the First Workshop on IoT-enabled Healthcare and Wellness Technologies and Systems* (pp. 17–18).

Troitzsch, K. G., & Gilbert, N. (2005). *Simulation for the social scientist.* Open University Press.

Turing, A. (1937). On computable numbers, with an application to the Entscheidungsproblem. *Proceedings of the London Mathematical Society, 4*, 230–265.

Turing, A. (1950). Computing machinery and intelligence. *Mind, 59*(236), 433–460.

Von Neumann, J., & Morgenstern, O. (1953). *Theory of games and economic behavior.* Princeton University Press.

Watkins, C. (1989). *Learning with delayed rewards.* PhD thesis, Cambridge University, Cambridge, UK.

Weichhart, G., Stary, C., & Vernadat, F. (2018). Enterprise modelling for interoperable and knowledge-based enterprises. *International Journal of Production Research, 56*(8), 2818–2840.

Weizenbaum, J. (1976). *Computer power and human reason.* W.H. Freeman and Company.

Wen, Z., Yang, R., Garraghan, P., Lin, T., Xu, J., & Rovatsos, M. (2017). Fog orchestration for internet of things services. *IEEE Internet Computing, 21*(2), 16–24.

Zhiyuan, J. L., Jung, J., Goel, S., & Skeem, J. (2020). The limits of human predictions of recidivism. *Science. Advances, 6*(7), eaaz0652. https://doi.org/10.1126/sciadv.aaz0652

Zhu, R., Lin, J., Becerik-Gerber, B., & Li, N. (2020). Human-building-emergency interactions and their impact on emergency response performance: A review of the state of the art. *Safety Science, 127*, 104691. https://doi.org/10.1016/j.ssci.2020.104691

Zhuang, C. B., Liu, J. H., Xiong, H., Ding, X., Liu, S., & Weng, G. (2017). Connotation, architecture and trends of product digital twin. *Computer Integrated Manufacturing Systems, 23*(4), 753–768.

Printed in the United States
by Baker & Taylor Publisher Services